Foreword by B

Paul Whitehouse
& John Bailey

HOW WE
FISH

The Love, Life and Joy
of the Riverbank

MUDLARK

Mudlark
HarperCollins*Publishers*
1 London Bridge Street
London SE1 9GF

www.harpercollins.co.uk

HarperCollins*Publishers*
Macken House, 39/40 Mayor Street Upper
Dublin 1, D01 C9W8, Ireland

First published by Mudlark 2023
This edition published 2024

1 3 5 7 9 10 8 6 4 2

A catalogue record of this book is
available from the British Library

ISBN 978-0-00-855967-0

Printed and bound in the UK using 100%
renewable electricity at CPI Group (UK) Ltd

CONTENTS

ABOUT THE AUTHORS

Paul Whitehouse is a comedian, actor and co-star of the hit BBC Two series *Mortimer & Whitehouse: Gone Fishing*, alongside his old friend Bob Mortimer. He also co-created and starred in the *Fast Show* and in many series with Harry Enfield.

John Bailey taught himself to fish aged five and has been eradicating his mistakes ever since. After leaving teaching thirty years ago, John has made a living by guiding anglers all around the world, writing and broadcasting about fish and fish conservation, and, of course, working on *Mortimer & Whitehouse: Gone Fishing* these past six years. Game, coarse or sea fishing; bait, lure or fly fishing: John loves every branch of the sport equally and reckons he enjoys just about the dream life.

ABOUT THE ILLUSTRATOR

Carys Reilly-Whitehouse is an illustrator and visual artist living in Glasgow. She specialises in intricate and detailed ink pen drawings inspired by antique medical diagrams. Carys studied Woven Textiles at Winchester School of Art and then went on to the Royal College of Art, where she developed a practice spanning textiles and contemporary art. In addition to her illustration work she uses her visual art practice to advocate for better understanding of chronic illness and patients' experiences. She now divides her time between working, exhibiting, music, and spending time with her partner and cats.

FOREWORD

BY BOB MORTIMER

You don't meet many extraordinary people in life, but I have been lucky enough to share a friendship with two of them: Messrs Bailey and Whitehouse, the authors of this superb book. Over these last six years they have both been teaching me how to catch fish in our rivers and lakes, and now they have decided to do the same for you. I recommend you listen very carefully to their advice. I always did and now I have a thirty-pound pike to my name.*

If you are already part of the angling fraternity then you will be very familiar with the exploits and writings of John. He is quite simply a legend, a slightly bow-legged one at the very top of his chosen profession. I too have slight bowing of the legs and have been told on numerous occasions by my osteopath that this is a feature of the classic athlete's body. No surprise, then, that in his teenage years John was a semi-pro footballer with teams like Stockport County in the lower

---

* Twenty-eight pounds, according to John and Paul

divisions. And always a reserve, he modestly adds! His life-time passion, however, has always being angling and the passing on of his knowledge of everything piscatorial.

To understand John and truly appreciate his gifts, you need to watch him from a distance as he surveys and inspects the beat or swim that we are about to fish. He stands like a heron, deadly still, as he inspects the water and assesses its promise and its failings. John never tires of telling me that this is the most important part of the day's work and to neglect it is to invite failure. It's tough advice for me, for as soon as I set myself down by the river I just want to lob my bait in, start fishing and commence the relaxation. (John will often let me do this because he's a lovely man and he wants me to enjoy myself.)

What exactly John is looking for remains something of a mystery to me. He's looking at flows and swirls and eddies and back currents and pools and all sorts of fascinating watery movements. He looks at the heavens above and considers the direction of the wind. He might lift up a rock or two and see what little critters are lurking there as a guide to which fly or bait to use. At the end of his silent musing he will walk over to me, point to a spot on the river or lake and tell me to cast at it. This will always be exactly where the fish we are targeting are lurking. John would catch quickly. I might need a day or two. If and when you next fish you probably won't have John with you, so my advice would be to pay close attention to all mentions of watercraft within the book.

John is, of course, the fishing consultant for the *Gone Fishing* TV show. When it comes to where we fish and how we fish, he is the master of that domain. Paul and I may chip in with the suggestion of an area we would like to visit, and I will always suggest that we once again go in search of a large perch (still haven't got one), but John will shape and design the fishing side of the episode.

When we arrive at the riverbank John and Paul will always form a little huddle to discuss tactics and tackle and whatever else serious fishermen talk about prior to the first cast of the day. Proper anglers love these chats, a great opportunity to display their knowledge, learn off each other and display their latest bit of kit. John will already have prepared my rod, so I just cast away, desperate to catch something before the experts do. As soon as we start filming John watches me like a hawk, and whenever there is an opportunity pops down and tells me what I'm doing right and what I'm doing wrong. If I'm float fishing he will also give me a tally of the number of bites I have missed. He has the patience of a saint.

John likes to keep things simple. A good example is when he takes us fishing for barbel and he just attaches a big lump of luncheon meat to the hook and casts it in for the fish to find. For me the simplest techniques are the most rewarding, and you will find them all explained here in this book (I still haven't cracked the art of fly fishing despite hours of tutorials with Paul). I have started taking my eldest son fishing at my local carp fishery. The lake is peppered with anglers with hundreds of pounds' worth of gear and electronic nonsense, fishing with mysterious synthetic and flavoured baits. We turn up with a couple of cheap rods and a small tackle box, and do just as well as the equipment blokes by hanging a piece of corn under a float. It's all thanks to John and his patient teaching of the simplest methods.

Paul is my fishing buddy and lifelong friend. It was Paul that dragged me off my sofa following my heart operation and took me fishing on the River Test. I wasn't immediately hooked on fly fishing, but I was completely won over by the idea of spending six hours by the river with this wonderful, beautiful man. As we get older it seems we spend less and less real quality time with our mates, and it's a crying shame. When I'm out fishing with Paul we put the world to rights, unload all our burdens on each other and, above all, laugh like drains. It's the fishing that gets us out and about, but we never let it get in the way of having an excellent time.

Paul loves his fly fishing and I always enjoy taking time out to watch him casting away with dogged determination and a big dollop of skill. Paul doesn't claim to be the greatest fly rod operator in the game, but believe me, he's caught a lot more fish than some people who do. *Gone Fishing* has forced Paul to rediscover some of the skills he learned as a young lad fishing on the canals of London and the banks of the River Lea. I think it's a journey that he has really enjoyed. He might be an old hand, but the look of pure joy on his face when he caught that giant barbel on the River Mole or that huge sea bass in the English Channel showed that the young platform-heeled Paul is still in there somewhere.

Paul is incredibly patient and giving towards me when it comes to the actual fishing. Like any newbie I get into a lot of tangles and lose a lot of his equipment. He never has a harsh word for me and always encourages me when I get frustrated or feel like I'm ruining his chances of catching. He does hate it when I start winding when playing a fish, though. 'Don't wind!' he will shout as I attempt to resist the urge to do exactly that. He's trying to teach me to do things right, that's all. He wants me to be a better fisherman because that would make us both happy.

Whether you are just starting out fishing or maybe returning after a long absence, I hope you have a friend like Paul who can accompany you on your journey. From the River Test to the River Tees, there are some breathtaking places to enjoy. If it wasn't for fishing I would never have discovered the majesty and beauty of the River Wye or the peace and tranquillity of the River Barle. Every single trip is a new adventure, a fresh opportunity to dissolve oneself into a wondrous environment. I am very grateful to have discovered the world of the angler.

Thank you, John.
Thank you, Paul.

A crucian carp

INTRODUCTION

Paul Whitehouse: I wouldn't say that I turn to fishing when I'm in trouble exactly, or even when I'm especially happy, but rather it has been a constant in my life, always there for me whatever. There have been times when I have been completely obsessed by it, but more often it's been a refuge for me, a happy place in my mind that I can always access when I need to.

I've led – and I'm leading – a fast-paced sort of life, and my mind can race away with me to the point where I'm plagued with insomnia. Many is the night I wake up at 3 a.m. with my thoughts cascading all over the place, and I'll lie there with some tune going round and round in my head. 'Stop that, Whitey,' I'll say, and that's where this blessed sport called fishing comes in.

In my mind I'm up on the River Dee, the river that's been my hallowed place for quarter of a century. I'll clear my head of worries, songs and everything else, and I'll fish that pool down. There I am, physically in my bed but four hundred miles away mentally, executing a single Spey cast with my line pinging out like an arrow, in that comforting salmon fisher's rhythm: cast and move, cast and move.

I never catch anything on these nights – I hardly ever do anyway – but that's for the best. I wouldn't want the excitement to disturb the harmony of my vision. It's a glorious oblivion at odds with reality, where there's never wind, rain or boiling heat, and I never get snagged or lose a fly. It's just my perfect session, tucked away in my mind, that I can reach for whenever I'm in need. My head – my health, even – would be in a different place but for my fishing; that's how central it is to me. I hope this book can get at least some of that across and convince any readers who don't already know what fishing can do for you to get cracking.

John Bailey: How We Fish *takes the techniques used in* Gone Fishing *and fills in the gaps that TV necessarily creates. It also explores more deeply what the sport is about and delves into the reasons we love it so very much. It is a testament to angling – a love letter, if you will – or at least a 'thank you' for the joy it has given the both of us.*

Paul Whitehouse is the most genuine man I know. When this book was proposed Paul was seriously concerned that he might be seen as 'cashing in' on the show's success. There are those in the public eye who might not care too much about the ethics of what they do, but Paul does. Deeply. I had to convince him of my belief in this book, my heartfelt desire to explain what fishing means to us both and, indeed, to examine how we fish.

I'd also like this book to be a way of us giving back, of setting others on the path that has brought us each a lifetime of the most intense pleasure and satisfaction. Merrily, merrily down the stream we go. So why not get on board with these tales of our angling lives?

Perhaps, in part, because of the success of *Gone Fishing*, the last few years have seen a sharp rise in the number of those who would like to go fishing or have wanted to return to a sport they enjoyed as kids. The trouble is that fishing isn't at all easy to access or to make a start on.

Go into any tackle shop and the array of gear is frightening, especially now that megastores are the norm and the small-time dealer, who probably sold tobacco and pet food alongside crow quill floats, has long gone. What is lacking today is the old guys (normally men but not always – there were, of course, some legendary tackle-shop ladies) dispensing advice along with the maggots and hooks and floats. Many of them showed endless patience and regarded tackle dealing as a calling, a sacred duty to pass on knowledge from one generation to the next. John, for example, learned to tie his beloved half-blood leaning against a counter at the age of six and has never really seen the need to try much else in the knot department since. But there are further questions: Where to go? What do you need? How do you set up? All these basic steps are vital but hard to fathom in the modern, glossy tackle emporium – or even on the internet that so often flatters to deceive.

Catching a single fish can often seem an impossible dream, and that's where we hope to step in. We're the old boys who used to help the first faltering casts, the first hesitant sessions by the water. We've got over a hundred years' fishing experience between us, and we'll share the lot in this book. We'll share our ignorance, too, because fishing is something you never fully understand or have all the answers to. It's the one sport that constantly surprises, and it's that magic that keeps us both as mesmerised as we were the very first time we cast a line on the water.

We're all about simplicity as fishermen, cutting out the unnecessary clutter that today's tactics and the tackle industry have cooked up, often in the name of profit. We truly believe that with simplicity come depth and clarity. We're not saying that we are mired in a dusty angling past; more that we're down to earth about a sport that should be easy, accessible and stripped of gobbledegook. We've both absorbed the advances made over the last half century; but we've not thrown out what worked before, and we haven't forgotten a lot of older lessons and skills that have been successful for ever.

Some of what we learned in our short-pants days has stayed with us, and what we knew way back then still forms the basis of how we think and fish today. The funny thing is that while tackle and bait have both made huge strides within our angling lifetime, a good deal of the really interesting, tactile stuff that makes fishing an art form has been over-looked these past forty years. We'll talk about the best of the modern and rescue the best from the past. At least that's our aim.

We're all about diversity, too, breaking artificial barriers between the different disciplines within fishing. You can trace these back to Victorian England for sure, a time when the mill owners fished the pure salmon and trout rivers of the country and the mill workers caught what they could in the canals close to home. We've never been fans of class division. If any technique is fun and exciting, and if any fish is worth catching, then we'll be up for the experience: dry fly on a chalk stream; floating crust on a carp pool. It's all the glorious sport of fishing to us, and we'll go posh or down and dirty as the mood takes us and as the opportunities arise. We simply don't hold with the one method/one species mentality. There's so much variety and so much to enjoy in angling that it would be a massive shame to deny yourself what's out there. That's the philosophy in *Gone Fishing* – and that's what you'll find here in *How We Fish*.

We realise our luck and appreciate what a privilege it is to write a book about what we love. And perhaps we are writing it at just the right time for us. We are not saying we are old enough to be on the edge of existence exactly, but with age come experience and wisdom, and sometimes sufficient knowledge to accept that you don't know everything and never will. That's certainly true about fishing.

There was a night nearly seven years ago when we were sitting by the River Wye, sharing a single rod. It wasn't dark yet but it was getting there, and the early bats were skimming the water. It was comfortingly quiet, too, apart from the flick of a few small trout and the occasional sploosh of a frustrated salmon marooned in the tail end of a low-level summer pool. It was so tranquil. We weren't talking much – when we did it was in hushed tones. It was almost like being in a church or some other holy place where you just drink in serenity and peace. Or those lulls in your life, just before sleep, when you're happy, secure, in a good place.

And then, crash, bang, wallop! Out of nowhere Paul's rod is hooped in his hand, his reel is screaming and our world has gone crazy. We're hopping around like the kids we still are, scrambling to control the fish, put up the net, and trying but failing to bring order out of chaos. Then there's a barbel lying in front of us in the half-light. It is massive and gorgeous with its burnished bronze scales and coral fins. It swims away, gold in the torch light, and we are as happy as grown-ups can be. As good as it gets. Ecstatic in the moment.

We met years back, when the TV show first started, though our individual paths had skirted a couple of times before then. What soon became apparent was that as anglers, we're birds of a feather. Paul is slightly more fly focused than (so-called) Bait Fishing Bailey, but we both cover all the fishing bases, and each of us even won fishing matches back in very much younger days.

What's been wonderful through years of filming is the obvious fact that we look at water and fishing challenges through similar eyes. We read what's going on and communicate almost telepathically. We take enormous dollops of pleasure from fishing, from trying to catch anything in every way possible, providing that way is simple, thrilling and moderately efficient, too. The joy of fishing and how to do it. That's entirely what we want you to get out of this book.

PART ONE

WHY WE FISH

CHAPTER ONE

WHY WE FISH

Paul: Blimey, there was a time when we never thought to justify our fishing or even explain our love for it to ourselves or anybody else, come to that. Everyone fished; it was something we all did. Half the kids in my primary school class had rods and every family had at least some tackle, even if it was stuffed in the attic. Angling was an unquestioned part of national life, like doing the pools, going down the pub or heading to the seaside on a bank holiday. The rich went salmon fishing and the poor went to a pit or a pond – I went to Kings Weir on the Lea – but we all went somewhere without a thought of doing anything else, outside of football and cricket, of course. It wasn't a case of why we went fishing. It was a case of why ever wouldn't you?

John: What I do know, partly from being a history teacher who ran fishing clubs, is that you can't push anyone into being an angler. The passion is either in you or it isn't. That's the tragedy of modern-day childhoods for me. The past thirty or forty years, kids have lost their freedoms and have to do everything now under strict adult supervision of one sort or another. Children can't know if they like fishing or not because they never have the opportunity to explore it properly on their own.

For me and Paul and most anglers over forty years old, fishing was our private existence of magic and mystery. It was our entry to the natural world and we went on our tod or with like-minded mates. We faced our challenges, solved our problems, confronted our fears and came out the other side stronger. Kids are so divorced from the outdoors now that the most recent Oxford Junior Dictionary *omitted words like acorn, adder, catkin and conker! Paul and I were deeply embedded in nature as children, and angling was simply an expression of that. If this book can get anyone, kids or adults, back to fishing, back to cold, wet weather, back to nature, then that's our whole aim.*

Why do we fish? We can rattle on for a long time about fishing being in our DNA, about its rich literature, about tackle collecting, fly tying, its defeat of stress and a hundred other aspects that are a part of fishing but not at its core. Fishing is truly about immediate, intense memories that imprint themselves for ever.

Paul remembers his first ever humble roach from the humble Lea Navigation on a humble shred of Mother's Pride white bread. It was all ever so humble when he was uprooted from the Rhondda Valley to Enfield at the age of five and this summery afternoon saw his fishing beginnings. Dad was leader of the Whitehouse adventure, Mum and sister spectators, and Paul was the wide-eyed pupil. It began with a float, a delicate crow quill, tipped with a hat of vivid scarlet paint the colour of a look-at-me smear of lipstick. Paul was handed the rod and told to watch that speck of colour and never, ever let his eyes stray from it. He was captivated, mesmerised, and when the float finally went under, the Whitehouse family shrieked out in unison, 'It's gone!'

This was magic, real magic. This was no card trick or third-rate subterfuge from the end-of-the-pier show. That float was this lad's connection with the underwater world and its disappearance created a joyful astonishment that burns as brightly now as it did then. The young Whitey looked at that roach, all four inches of it, and he knew he had never seen sheer beauty before. The silver-and-blue tinged flanks, the scarlet-and-black centred eye, and the fins of burnished red left him agog with a wonder that has never palled. Every capture, even now, conjures the same miracle, illuminates a similar memory.

For John, too, a red-tipped float was his object of wonder for five years, till he was eleven and lost it one wild, wet, windy morning when a Norfolk tench stormed into a reed bed, taking the tackle with him. The loss of the fish was bad; the demise of the float was worse. That concoction of crow quill and cork body had been his constant waterside companion.

One evening in particular is eternal, at dusk, always the magic hour, when the street lights were on and the old mill dam shone in pallid mirrors of watery gold. The quill rose, lay flat, moved a foot to the left, disappeared beneath the surface. The rod, a brutish tool of solid bamboo, grunted as a proper perch bored this way and that. That fish was lost, but it was seen three times, rolling just out of reach, huge, barred in black stripes, a dream-like monster to a kid of barely seven years old. 'You can't lose what you never had' goes the wise old saying, but try telling that to John that July evening as his tears streamed down and Dad through gritted teeth told him to shut up, buck up and let's get home.

So, for us, don't try to pin fishing down to this, that or the other. Fishing's more of a life force, a desire that can kindle when you are a kid and will burn till you die at whatever age you might achieve. It can also be a life saver. John had never taken the issue of mental welfare that seriously, being an old-school northerner at heart. You could be sad, you might be down in the dumps, but 'pull yourself together, lad, and get on with it' was the dictum he had grown up with and lived by.

John: Not quite as easy as I thought, this life of ours. Winter 2021/22 was a very bleak time for me. Though I hesitated to admit my gnawing despair to any but a very close few, I realised I was experiencing mental depression of a kind I had spent a life hearing about in others but doubting in myself. Deep down in my mind and, worse, in my soul, I felt a different person, like some volatile beast was eating me from the guts out. Everything became difficult, then threatening and finally terrifying. Waking up in the grey dawn light became a horror for me, and by the start of December, I was at my wit's end. My two pillars in life, my wife Enoka and Paul, worried hugely – and simply told me to go fishing.

Throughout the winter and into the first stirrings of spring, I left the house at around 2 p.m. and drove for fifteen minutes through the placid Herefordshire countryside to a swim on the Wye. (For the dead keen fishers out there, I chose this swim because it was possible to access in the winter muds, because it fished well in most river levels and because there were few serious snags.) It became my place of calm, of salvation. Night after night, I nestled there amid the reeds, watching the sun set over the hills directly opposite. I listened for the stable hands to feed the horses in the paddocks, waited for the geese to flock in, for a shower of goldcrests to swarm into the alders and for my very own pal, the robin, to come and visit. At 5 p.m. every day, the front porch of the black and white cottage on the floodplain to the west was lit, and around the same time, on clear nights, the stars began to sparkle in the black, Welsh/English borderland skies above. It was a time of peace, reflection, deep breaths and increasing wonder at the majesty of the natural world. A few fish came my way – brilliant, battling barbel that quickened the blood – but most nights passed quietly. It was akin to a hibernation.

By April, I was a man restored. Better than counselling. Better than pills. Cheaper than both. The love of a wife and friends and the healing powers of that swim on the dark, deep, ever-rolling River Wye had done it all.

So, you can see the depth of meaning that fishing gives our lives, but we also like to think that we give back. In our view, it is an incontrovertible truth that no one out there in the field of conservation cares for fish in the way that thinking anglers do. This is not self-interest – two anglers simply wanting to magic up more or bigger fish so that we can catch them. Not a bit of it. Fishing engenders love for fish, a passion for pure waters, and it is no coincidence that both of us are trustees of Fish in Need, a charity fighting for a better deal for our finned friends.

Anglers also understand rivers in a way no one else does. That's not some sort of idle boast, but a recognition that anglers spend more time by rivers, watching rivers and trying to work rivers out than any other group of people. This concentration leads to an absorption that can be exceptionally clear-sighted. Sometimes answers invisible to most people are in clear view to committed anglers. Their approach to issues on the river is elementary but almost always correct; false arguments are put aside, and simple truths are realised.

There is a philosophy that enjoying nature is good for the soul and we'd agree. Canoeists and swimmers are rightfully allowed to use our rivers and waterways, but an understanding of the delicate ecosystem that nurtures aquatic life should be rule number one before setting forth on a river. Not all anglers are saints, but river welfare is at the core of what we do and love. Anglers also pay for the privilege to fish. It's illegal to fish without a rod licence, and we often pay on top of that to fish a particular venue. This money goes to the Environment Agency or private owners, or to clubs and associations who lease waters, monitor them and work hard on their upkeep. For that reason alone, the question of why we fish is answered: for our own wellbeing, of course, but also for the wellbeing of the waterways we visit.

Indeed, without fisher folk, our nation's fish would find themselves in a far darker place. We've all got a part to play in this mission to make the voice of fish heard. Perhaps you join the Angling Trust or the Wild Trout Trust, both of which initiate essential river rehabilitation. Perhaps you engage in group-working get-togethers organised by your angling club and improve river habitat for fish welfare. Local Wildlife Trusts, even the National Trust, all show remarkably little interest in fish or concern for their welfare. Positive pro-fish input from all of us could change that in time if we worked together. There are many anglers who never get off their backsides in this respect. Let's be different and fight for our fish when and where we can.

John: When I was a kid, I lived a woodland-walk away from the River Goyt, a tributary of the Mersey, in Greater Manchester. Aged six, I'd trek down to Otterspool, a glorious water – a churning place that had lived up to its name sometime before the Industrial Revolution, but in the very late 1950s ran orange or green or whatever dye the mills upstream were using that day. I loved the roar of the weir and the scent of lushly damp loosestrife, but what I'd have given for a fish or two. Today there are more than a fish or two – there are lots and lots. There are even barbel and grayling, two species, like dowager duchesses, disdainful of any home that isn't the finest. Miracles can happen.

There was and still is a pub close by, which my dad was in the habit of frequenting on warm summer evenings. One August, when the garden was all abuzz with the insects of the night, I learned a big lesson.

Two United States Air Force lads were staring into pints and talking about their fishing. It seemed they hadn't had a bite in days, and just as I was about to squeak up and put them straight about the fishless waters of the Goyt, Dad caught my arm.

'Leave it,' he hissed. 'Don't spoil their fun, eh? It's their holiday and if they're happy, that's the thing.'

'Don't worry, sir, we know,' said the older of the two, who had the hearing of a bat. 'We're just having a fine time, fish or not.' He looked at me and could have repeated the old mantra that it's not called catching, it's called fishing. But he didn't. 'A fine time,' he repeated. 'Who needs fish to spoil things?'

CHAPTER TWO

WHEN WE FISH

We anglers are so lucky that our sport is 24/7, that you can dangle a line pretty much every hour of any year for something that swims and has fins. Even as we write these words, there will perhaps be an Inuit in Greenland pirking for a great halibut in frozen water a thousand feet deep, a polar bear looking on with hungry interest. Or two friends on holiday from the Indian plains casting a fly for a Himalayan brown trout, Everest resplendent way to the north. In New Zealand an angler might be tussling with a mighty rainbow trout in the crashing waters of the Tongariro River. Someone, somewhere will be fishing this very second. It's the sport that never sleeps.

We fishers are also blessed with the element we most commonly explore. Whether our fishing is done in running, still or salt water, what we see is beauty, a gorgeousness set ablaze by the sun's rises and sets, times when it might seem that you are casting into a cauldron of fire. How kind of the fish to feed hardest early and late, even if it means that you need to buy a good, loud alarm clock and be prepared to take your supper close to midnight. Not that the heat of the day should always be overlooked. Some mighty fish have been taken in baking conditions, when the watery world has been moribund under the noon sun. But we want to talk about night, a time when few practitioners of their sport except anglers are about their business.

If we do go out after dark in this century, chances are that we'll be in a city, under neon, the sky invisible above us. Fishing at night in the deep countryside is a different deal altogether, winter and summer alike. Those pinpricks shining like diamonds in the black dome above you are stars, celestial creations we don't now see in the polluted light of civilisation. There's little in the way of silence at 1 a.m. by a woodland lake echoing to a cacophony of owls, the scream of a lonely vixen, the snuffles and snores of a hedgehog, and the shuffling badger digging the earth for worms and shrews. There are smells, too, that are new, curious and unforgettable. The damp soil of a molehill newly turned; the fragrance of the honey-suckle that blossoms in the dark; and the smell of water itself, heavy and dank with pondweed.

John, in his teens and twenties, fished almost exclusively during the hours of darkness because as a single student and then teacher it fitted his lifestyle. He found that if you fish a lot nocturnally, your night vision develops and you learn the sounds of the night and make sense of this parallel universe. A knowledge of the night and an easy familiarity with it gives life a new and deeper dimension.

And what of the fish? Unless you find yourself on a pool stocked with trout or carp delivered yesterday on a lorry, you are hunting a creature that is wild and hardwired for survival, one that knows darkness or subdued light affords it the cover it needs to feed unthreatened. It is in the darkness that most sea trout run up knee-deep Devon rivers. It is when the moon is up that shoals of big bream set out on their patrol routes, feeding hard until the farm cock crows. It is on a night tide that sea bass run into water barely covering their backs, and when conger eels slither from their rock caves to hunt.

So, if you are setting out to fish, at least consider being on the water outside of supermarket hours. Waters are wonderful, magical places, but treat them always with respect and, initially at least, with a dash of caution. Fish with a mate, the first few times at least. Take a torch, a phone and plenty of food (though remember to use your torch as little as possible, as the artificial light destroys the night vision you'll be acquiring). Wear warm clothes; even a mellow dusk can turn bitter by 2 a.m., the coldest part of any night. Make sure to tell family where you are going, when you'll be back and, perhaps for starters at least, fish a water where other night owl anglers can offer support if needed. Things can go wrong, like the time a twelve-year-old John poured his scalding soup onto a cup held unwittingly upside down in the darkness. Eight hours is a long time for a kid to sit alone with a scorched hand smelling of oxtail.

Weather is what fishing is all about. Wild fish do not eat all the time. Conditions create windows when they feed hard but then they can switch off again entirely. Just because you are not getting bites does not mean you are always fishing badly. No one can force a fish to feed if the weather is just wrong. Trouble is, it's often infuriatingly difficult to interpret weather patterns and how they may affect the fish. Educated guesses are the best we can make till fish learn to talk. By and large, settled conditions are good because fish adapt to them, become comfortable and feel happy enough to feed. But sometimes a change in air pressure, say, can enliven fish and spark them into feeding frenzies. Or clouds come over the sun and action starts. Or, conversely, the sun comes out, the water warms and fish begin to feed. You see how maddening the whole game is? There are few hard and fast rules and you can do no more than keep aware of every little flip in the weather, how the fish are reacting, and eventually you just might build up a hazy picture of what is going on in an environment that, to us, will always be tantalisingly impossible to comprehend.

One thing we are sure about is wind direction. South and west are good – in the UK, anyway. North is poor. East is a disaster, a complete turn-off for almost any fish that swims. Why this is so extreme is anyone's guess, but all water activity is suppressed and all fly life is completely subdued when the easterlies blow. Everything sulks in an easterly, but it sparks into life when the wind swings round to a blessed south-westerly. It's a 'light switch' moment, and a dead water one minute is on fire the next. It takes our breath away, even now.

There was a time in *Gone Fishing*, way back in series one, when we could not catch a tench whatever we did. For a day and a half the lake had been a tomb for Bob and Paul, and the camera crew were starting their slow wind down towards the moment when the director calls it a wrap. Then Paul's ears perked up … a rolling tench! Some flies, little black gnats, freckled the water. A fizzle of tiny bubbles burst close to his float, and he was like a cat waiting to pounce. Yes! Tench on. Quick, scramble, get those cameras rolling, get that drone back in the sky. A miracle tench, alright – all down to a slight southerly shift in the breeze of which Paul made instinctive and instant use.

When you have one of these windows, use it! Once again, back in *Gone Fishing* days, we had a bonanza of big carp feeding like starving pigs on a lake neighbouring the one where Paul and Bob were filming. Like a troop of rowdy hippos, they were muddying the water and gobbling up sweetcorn as fast as John could pile it in. Paul pleaded that he and Bob should try to fish for them, but the schedule wouldn't allow it. John was tasked to keep the fish feeding through that night, and he tried his best, feeding till 1 a.m. and again at a 5 o'clock dawn. But he knew it was in vain. Sure enough, just as the forecast had warned, at 2 a.m. the wind shifted east, the temperature fell 5 degrees and when the light broke the carp had gone, every last one of them. We struggled in that programme to catch one average tench. We could have had a thirty-pound carp if we had made the most of that extraordinary opportunity.

CHAPTER THREE

WHERE WE FISH

Paul: When I've got to get out of London, that's often for Gone Fishing, *and if I don't need a mountain of gear, I'll let the train take the strain. I'll get myself a window seat where I can indulge my own personal water radar I've got going on. It's some sort of sixth sense that makes me put down any script I'm learning or my copy of* Trout & Salmon, *or more likely the* Beano, *and have a look. Hardly ever do I see anyone fish a canal or a river that I'm travelling past. Ponds, though, they're a different matter. Everywhere I'll see anglers sitting behind buzzers and behind fences to keep out the otters. All the old locations which we fished as kids are ignored, forgotten about, and natural venues seem to have given way completely to artificial ones. Perhaps you can see the reason in this shift in location. On a commercial venue you know there are plenty of fish. You know predators, paddlers, wild swimmers, dog walkers and the rest are all kept out, and you have a guaranteed session of peace. It's less stress all round, which I completely understand.*

John: Yes, I see the point you are making, mate, of course I do. But please think of the natural waters we have fished together over the years and remember back to the times we have gawped at their beauty. Think of the Dee, which you love. Think of a pewter dawn, when you have been desperate to get out early and connect with one of those salmon that have moved in the night and that might still come up to investigate a fly. The silence broken only by the murmur of the water over stones and shingle. The mist over the meadows. Then the dawn chorus, tentative at first and only later breaking into full-bloodied song. A salmon swirls close to your fly, and without warning that loop of line draws tight in your hand and you're in! How can a concrete pond compete with that?

And me? When you told me what you see out of your train windows, experiences which are ever more the norm I know, I thought at once of India and the Kengal Rapids on the Cauvery River. I remembered nights sitting out midstream with Bola, my guide and friend. 'Fish are close,' he'd hiss, and I'd know it, feel it, almost dread it. I'd want to hook a great golden fish that could easily pull my arm off, but deep down the coward in me wanted NOT to hook one at any price. How can it have come to this? I wondered as I sat there. The lad who only dreamed of roach and perch from a northern canal is perched here now above half a mile of volcanic water, within touching distance of a demonic fish that is his own size but has twice as much strength, heart and spirit. What arrogance brought me here? I asked as I waited for the rod to be almost torn from my grasp. What did Kipling say about these fish, the mahseer? Something about the fact that beside them, the mighty tarpon is but a

24

herring? I've done that and more and have lived to tell the tale.

Paul, tell me, how can we compare what we have known to a pond fenced to keep all danger out? Aren't we selling angling and ourselves short in some big way?

Of course, we know this book isn't about exotic fishing, and we need to keep locations realistic and accessible, but there's a point worth making. If all we do is fish artificial locations where nature is kept out and the fish, be they trout or carp, have arrived by bus, then what is to become of the sport that is centuries, even millennia old? Life has become hideously sanitised, and to turn fishing into a health and safety exercise seems to undermine all that it stands for. If it is just guaranteed numbers of fish you are wanting, isn't this akin to dipping your line into a piscatorial supermarket where skill, jeopardy and challenge are reduced to an absolute minimum?

Many people have commented on Bob's explosion of excitement whenever a fish is landed on *Gone Fishing*. It's not put on for the show. It's a genuine tsunami of relief, delight and wonderment, all mixed together. In the show, Paul and Bob only ever try to fish for wild fish in places where these creatures have been born and brought up, where they have learned to survive against all odds. This makes them worthy opponents – quarry that actually hold the aces in most hands that we play. Fooling wild fish like these is an enormous challenge, and success cannot ever be taken for granted. Nor should it ever be. The fish might win. You might win. The outcome is constantly in doubt. It's the unknown final chapter that makes us come back over and over again.

It's probably true that commercial fisheries fulfil a need in modern society, but they shouldn't be seen as the ultimate, the be all and end all. Natural fisheries, by contrast, are places of wonder and we'd plead with you to sample them whenever you can. This does not necessarily mean that you have to go miles out into the countryside or gain access to some private estate where no one has cast a line since Victorian times. Urban fishing has come on leaps and bounds this century. The Wild Trout Trust has worked miracles in developing brown trout fishing in the most industrial of landscapes. A young hero of ours, Alan Blair, has shown on a hundred YouTube films the calibre of carp to be found in park lakes, in canals and rivers decorated with supermarket trolleys. There are serious numbers of fish back in the tidal Thames, the Trent has never fished better, and all you need is the imagination to find places of your own close to home. Check out the clubs in your local area; a good tackle shop will advise. Many will have an array of venues and might have fish-ins, social nights and run matches, where you will be welcomed into a true fishing family. Both of us started out as club anglers, and today John is proud to carry a Herefordshire and District card that offers unbelievable fishing for hundreds of accomplished River Wye anglers to enjoy.

What Paul observes from the train window is not strikingly new in one way. Since the late twentieth century many fishing locations have been engineered, but many of them always were. John was brought up fishing northern canals, which were designed by the engineer James Brindley and dug out by thousands of so-called navvies. Then John became a wee expert on Norfolk estate lakes, once again dug with shovels and picks for the gentry of the eighteenth century. In later life, he fished for tench in pits dug for the sand and gravel used in the construction of roads and houses – as well as runways during the Second World War. In fact, if it hadn't been for the bravery of the opposition to the Third Reich, tench fishing wouldn't be nearly as good as it is today through great swathes of the south-east of the country.

Where you fish is not always up to you, anyway. Geography plays its part and so does your age – passing a driving test lets you cast way further than ever before. For John, getting that licence opened up fishing for him in a way he'd never experienced before. For Paul, not so much. His bond with his car-driving dad meant that his fishing world already extended to the Usk, the Wye, the Thames, the Ouse and even the rivers of Ireland. That was a wonderful bonus. There is no better way into a fishing life than with a family member by your side.

Time, money and the amount of blood, sweat and tears you put into your sport are now the limiting factors, rather than how far your bike or the bus can take you. The aforementioned cliché about what we do being called 'fishing' rather than 'catching' does have resonance here. In our minds, where you fish is central to the whole process, and we are lucky to live in the UK, which has waters that are wondrously beautiful and even spiritually inspiring. Finding fish doesn't necessarily matter.

Just look at some of the *Gone Fishing* locations. Remember, for instance, the breathtaking park and ancient trees at Burghley, where Paul and Bob caught tench of mahogany magnificence. Or the Exe at Cove, that whispering stream with pools of incalculable depth where salmon sleep out the Devon day. Then there are the mighty, rolling Tweed and Tay, and the slip-sliding mystery of the Lea. North Uist raked by wind, sun and tide, and, as ever, the Wye, a majestic border between England and Wales.

And what about the intimate, tree-cloaked River Mole, still tiny where we fished, miles above its meeting with the Thames at Hampton Court? Who could forget the barbel that Paul and Bob caught, perhaps even fourteen pounds, which made the entire cast and crew gasp out loud at the sight of it?

There's a back story of great importance here. On the Monday (the fish was caught on the Wednesday) there was rain like the Indian monsoon in London, and Paul was caught for hours in water-drenched traffic. Down in Surrey, the Mole rose in the deluge, the sewers burst and the very swim John had been baiting ran chocolate with the untreated detritus from the town above. By Tuesday afternoon the Mole began to burrow back into its banks, and by Wednesday morning, when the attempt was to be filmed, the swim looked good, apart from stuff you wouldn't want to see on TV that John spent an unhappy hour prising out of the over-hanging trees.

The Mole looked terrific and that fish was a Moby Dick of a barbel, but let's not forget that almost every natural water that we fish is under threat one way or another. Unlike artificially bred fish stocked in a commercial water, naturally born fish swim a daily tightrope with dangers every side of them. Our rivers have simply never run more troubled courses.

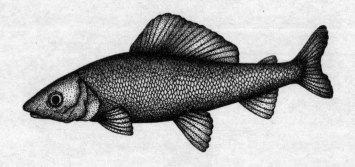

A grayling

CHAPTER FOUR

HOW WE FISH

John: It was Aristotle who said give me a child till the age of seven and I will show you the man. Though the concept has been used and abused by some over the centuries, when it comes to me and my fishing the great old Greek wasn't far wrong. There were two major influences on my piscatorial progress in the fifties. Mr Crabtree Goes Fishing *and the type of tackle that was available and within the limits of my pocket money.*

Let's take Crabtree *first – the cartoon strip drawn by Bernard Venables after the war and eventually brought together in a bestselling book. This was a volume that sold millions and was the bible of every young angler from 1948 until the eighties or even nineties. In* Crabtree, *Venables promoted an ethical view of angling that might appear old fashioned now but wasn't far wrong. Sharing the wonder of the sport. Appreciating the glory of nature. An emphasis on fish first, and tackle and bait second. And it was perfect in its simplicity. Everything* Crabtree *advised was straightforward, down to earth and practical.*

In the world Venables created, Crabtree and his son Peter caught fish because they understood fish; they studied watercraft and they had decent enough tactics and tackle to catch them as a result. Which brings me to tackle. Even if you had money back then, there wasn't that much to buy. Everything a northern lad owned could be bought for under a couple of quid and was basic stuff. All tackle worked but in a rudimentary fashion, and that first reel of mine could produce a tangle out of thin air, cast after frustrating cast. The rod was whole cane, and if I ever managed to put out a float more than a rod length out I was proudly astonished.

So, there you are, the twin factors that have guided my life in angling and out of it. Study your quarry. Provide yourself with the tools, but keep them simple and to the point. There's the saying: all the gear, no idea. It's a trap I suggest you avoid.

Paul: I'd echo most of what John says. What he did, I did, and nearly all kids did before the invention of bolt rigs, which meant that rods could then fish for themselves. For me and my gang it was Kings Weir and the Lea that taught us fishing. I'll talk some more about this later, but it was the train and then a mile and a half walk, a trot, really, and we were there. All the way we'd be thinking about the swims – the Hemp Swim, the Chub Swim – wondering which would be free, where we'd get to weave our magic. I had a key to unlock the gate there, which made me feel very grown up somehow and, best of all, gave me entry to a perfect world, the pristine Lea, the river of Izaak Walton, father of fishing.

It was a hard river even then, but full of quality fish. It demanded that you fish with stealth, that you used your water-

craft. And though we did use the occasional block-end feeder, this was never our favourite method. We were at our happiest flicking out meat, drifting floating crust, freelining lobworms, feeling for bites, keeping on the move. We'd look out for dead minnows and gudgeon, for slugs; we'd scavenge whatever we could and we'd get cracking chub – four pounders that looked colossal to us then, and we rarely got bigger than that. It was magic stuff because it was simple and direct and our way into understanding nature. I joke that fishing is made up of periods of boredom punctuated by short, sharp moments of intense boredom, but I don't really mean that. Absorbing the life of the river was what fishing was to us then and what it is to me even now.

Carp anglers often have a bad press nowadays, but not all of them sit and sleep in bivvies, oblivious to the lake outside the canvas. The best of them are brilliant: they arrive at the waterside, probably after a day at work, but they take their time. They might only have till 10 p.m. to fish, but they know that one perfect cast can do the job better than a hundred flung out blind. They walk slowly, they watch everything, generally through binoculars and polarised glasses. They look for fish, signs of feeding fish, weed growth, wind direction, stained water, bubbles, scatterings of small fish and a host of seemingly insignificant clues that would pass many anglers by. They strategise as they go: they work out not just where the carp are, but where best to fish from, where to hide up, what angle with which to execute that one spot-on cast. Only when they are absolutely sure of their plan do they move into position and put out a bait. The tackle is way down the list of their priorities; they assume it is up to the job and then they forget about it completely. Where the carp are, what they are feeding on – these are the complexities of the game, not the ironmongery on the end of the line.

One of the best fly fishers we know is also a guide, and she likes to get to the river around sunrise, in part so she can compose herself in the tranquillity of that hour, settling into the rhythm of the water and valley around. She might walk the bank, looking for otter tracks in the dew (if an animal has been around in the night she knows that the wild fish will be on red alert and super-hard to tempt). Above all, she is looking for the webs of spiders, illuminated by the steadily growing light. Each one she comes across she will study with an entomologist's eye. She'll analyse the terrestrial insect trapped there but will pay special attention to the waterborne species. Mayflies, olives, stone flies, anything and everything will help her establish what the trout might be focused on during the coming hours. She won't be fishing blind, putting her faith in a swish rod and reel. No way. She'll have got to the core of how the river is working that particular day in that particular season. Time spent watching is worth double or more actual fishing effort. It's called watercraft.

What is watercraft, that oft-used term that anglers have chewed over since Izaak Walton was a lad? It's what you don't need on a commercial fishery where every yard of bank is identical, the bed of the pond is as flat as an ironing board and where if a weed sprouts it is instantly rooted out. Watercraft is bonding with the river or lake or seashore you are about to fish. We cannot overstate the importance of all this. What you are looking at is a whole waterscape rammed with clues, and your imagination will unlock the puzzle only if you give yourself time. John especially would *never* go to a new water with fishing tackle on his first visit. He would always allow himself the time to walk, watch and ponder.

This is how it has worked for years with *Gone Fishing*. Let's take a certain glorious Shropshire crucian lake as an example. Before filming began, John visited three times, for a full day on each occasion, beginning at dawn. What he was doing was plotting where the crucians might be, where and when they might be feeding and what conditions would spark them into periods of activity. He looked for rolling fish, bubbling fish and those tiny clusters of bubbles that crucian carp create when they are sifting the bed of the lake. By the time Paul and Bob arrived with the crew, he had formed a pretty fair idea about where the fishing might take place successfully and why. As it happened, the weather didn't play ball, but the knowledge stored up during all those hours of apparent inactivity did in the end save the day. (Paul: *Sounds good, John, but it was actually me seeing a crucian roll, me casting to it and Bob getting the run when I went off to have a go with a float that saved the day. Eagle-eyed, quick-thinking Paul and goal-hanging Bob, if truth be told.*)

Or take the episode on the River Wye, close to the town of Ross. John arrived on 15 June, the day before the fishing season opened. He walked the entire stretch of river three times, binoculars at the ready, a notebook in his pocket. Above all, he listened to Adam, the owner of the beat and an acknowledged master on it. Local wisdom should always be sought, however good you might like to think you are. Adam talked about the beach, which was an obvious place to begin, but there was a pull towards the rapids a mile upstream that John found hard to resist. The water just looked right, and he worked hard to forge a way down the precipitous bank that the camera crew might use. The sacred date of 16 June dawned and the fishing on Adam's beach was hard indeed.

When filming finished, John took himself back up to the rapids to think things out. On the point of dusk, out in the quickest of the flow, a single barbel rolled in an easy but unmistakable porpoise-like movement. That one fish was enough. He was back before dawn the next day, a mere five hours later. Now he fished, bouncing large pieces of luncheon meat around the rapids from a point on the bank that the crew could reach safely. We say he fished: the important thing to note was that he tied the meat to the line without using a hook. He simply wanted to locate precisely where the barbel shoal might be without the commotion of hooking a fish, which would very probably ruin the filming chances. Every time the moving meat reached a certain dip in the river bed behind a large boulder, it was sucked free off the line. The precise place had been found, and at 9 a.m. Paul cast to it expertly. A hook was tied on the end of the line and a beefy barbel found itself fooled. One cast and job done, but only on the back of watercraft and hours of painstaking observation.

Above all things, we believe that the key to success in your angling life is not to spook the fish. So vital is this element of angling that it deserves a chapter all to itself.

CHAPTER FIVE

DON'T SPOOK THE FISH!

John: Harvey, the late, legendary and much-loved spaniel, was intent on watching a shoal of chub in clear water on a summer river. He lay quite still, his eyebrows barely showing over the top of the tall grasses between him and the fish. Then, calamity, the bank collapsed, pitching a black dog in among thirty fleeing fish. Harvey, to them, must have been the exact size and shape of an otter, and the group of fish bow waved over three miles down-river in sheer panic, never stopping, covering the whole distance in less than half an hour.

Wild fish have to be attuned to every shadow, every tremor of the water around them; in fact, their whole survival is a tale of success against the odds. These four- and five-pound fish Harvey disturbed were around ten years old, and they had arrived at that age despite fearsome dangers threatening them every day of their lives. We humans have no idea of the senses that keep fish safe, but to get close to catching them it is paramount that we understand their world and do everything possible not to unsettle it. Study to be quiet, said Izaak Walton hundreds of years ago, and never has better advice been given to the aspiring angler.

Paul: John's correct that we don't know how wild fish learn to survive, but you need to take the fact that they do so seriously. I can't count the times I have walked up a chalk stream very cautiously and still I've caused twenty trout to bow wave ahead of me like the devil has been after them.

Trout on rivers like the Test are supposed to be bomb proof, especially compared with highly strung Yorkshire river fish, but my experiences say they are not. It's sometimes like there's a collective fear among all the trout on a beat and that they are all in some sort of state of defensive vulnerability in unison. I compare them to those aerial displays around a starling roost, where countless thousands of birds dance to the exact same tune. Yes, there are some days when you just cannot get near any of them, and I have no answer for it.

Is it atmospheric, or the light, or the time of day? Have predators or poachers been on the river in the night and that's scared them and put them all on point, forever on the fin? We'll never understand fish more than partly, and that's a good thing, one of the great fascinations.

Scare the fish you are after and you are done for, with no hope whatsoever. A scared fish will rarely, if ever, feed, no matter how good your bait, fly, lure and tactics might be. Don't shoot yourself in the foot before a cast has even been made. These are Walton's Warnings, if you like. To be obeyed at all times.

It's dawn on the riverbank and look hard at the dew; or, if you are after pike or grayling in the winter, look hard at the frost. Do you see footprints of other anglers heading in the direction you want to take? If so, turn around and make new plans, because someone else has destroyed your chances before the sun is even up.

It doesn't have to be a human that scuppers your morning. An otter will sink your chances, too, so learn to recognise the signs they leave and give them the widest berth.

Paul: It's fine to talk about signs of otters, but what are they? Spraints are the big giveaway – droppings left to mark territory. These can be black or grey, or even pinky coloured, often sausage-shaped and crinkly looking. Nice. They might contain fish bones or scales, or the remains of signal crayfish now that these buggers are in most watercourses. Their tracks are generally two or more inches wide, with five toes and short toenails that leave only slight impressions. Look for muddy slides where otters repeatedly come in and out of the water. You might even hear them whistling to each other if they are close by. And if you see one in the water, it's likely to be a glimpse of something smooth and black. If it sees you, it'll dive and leave a string of bubbles. I love otters. I've bumped into them eyeball to eyeball on many occasions, and they have lifted the day for me. But they can cause havoc in the fish world …

John: Otters can cause havoc, for sure, but wild-born fish soon build defences against them. They are hard wired for survival and show remarkable otter-avoidance skills. We have both fished at that cluster of waters we call the Snow White Lakes, and I remember a day when the bream were going barmy, rolling and clouding the bottom of this big lake. The bite indicators were never still and fish were coming thick and fast. Then, over a hundred yards across the water, an otter appeared, evidently having come from the adjacent river. It slipped into the lake and ALL bream activity ceased in an instant. The otter swam seventy yards to the island, where there was a kerfuffle and it reappeared with a coot in its mouth. (I should say 'he', for it was obviously a dog otter.) He swam back across the lake, got out from whence he had come, and within five minutes the bream were back on it like they'd never been away. HOW? How did they sense that lethal animal when it entered the water and was at least sixty yards from them? And how did they know equally instantly when the danger had passed? This incident is not unique in my experience, but it is a fine example of nature's survival mastery.

Between the years 2012 and 2021 I walked the River Wensum nigh on every day of my life. I saw endless otter activity of the kind Paul describes, but I only saw them unquestionably kill chub during and after periods of the most extreme cold. These prolonged spells of brutal weather blown in by easterly winds froze the chub into comatose blocks, which gave the warm-blooded otters the opportunity nearly always denied them. The angling brother- and sisterhood should take on board the fact that truly wild fish do not need fences to protect them.

Remember that any unnatural noise alerts fish instantly. Noise is a killer, and sound is intensified by a factor of five underwater, so lighten your footsteps and move like a heron.

Don't hurry. Take your time. Tread on grass or mud rather than gravel, which crunches as you arrive on the scene. Speak quietly or not at all.

Consider what you wear. While you don't need to make yourself look like a sniper, drab is the aim. A hat is good – anything to shield a pale face or shiny pate from the flash of the sun. Walk upstream so you approach fish from behind. Watch out for your shadow falling on the fish, and remember that it lengthens as the sun sinks. Don't point at a fish; it's rude and can do nothing but harm. Never scare small fry in the margins if you can avoid it – they'll scatter into deeper water and betray you irretrievably. Think mouse: creep, crawl and make yourself small.

Always make the first cast count in any fishing situation. Remember, the more you cast, the less likely you are to catch a fish unawares. If you are fly fishing, don't make more false casts than you need to. Learn to cast from a sitting position, or even lying on your side so trees are no longer a problem and your line can fly under the overhanging branches. Cast slowly and methodically, and think about your silhouette all the time you are fishing. If you are striving for distance, it is because you have driven the fish away from you, and the more you thrash the water the greater the spookiness becomes. You can end up a human windmill, flailing the air for no result.

Learn to cast on your back, crouching, even lying on your stomach. Aim to cast a fly, float or lure a foot above the water so it falls more lightly. If you are fly fishing, a long leader helps avoid putting the fly line itself over a wised-up trout, and if you rub that leader with dirt or clay, you'll avoid it glinting in sunlight. Never be impetuous, and if you are cast-ing out any type of altricial lure put a few casts down each margin before exploring further out – if you don't you'll scare the close-in fish, so tackle them first.

Look out for anything and everything happening on the water and use it to your advantage. If swans are upending and feeding on weed, get beneath them, because that's where fish will be, gorging on dislodged invertebrates. If cattle come down to drink, get close and put a bait immediately down-stream where the water will be coloured and fish will know to look for displaced worms.

When you are float fishing, hold back the float in the current so that the first thing the fish sees is the bait, not the shot coming to clip it on the nose. In crystal-clear shallow water, a wily carp can spook at the sight of a float. If distance is not an issue, tie on a goose feather or sliver of reed as an indicator. And remember, when you consider distance, accur-acy of cast is always more important than how far you can chuck a bait.

Bait, now there's a point to ponder. Fish are clever. Fish have memories. Catch a chub with a piece of meat on a ledger, say, and you'll not catch it again that way. Always assume that once a fish has made a mistake, it won't make the same one again.

Pike are a good example. If you have caught one on a sardine, it could very well be that you need to dye your sardines red if you want to catch that pike again. John took two fishermen, Ray and Adrian, on a guiding day some years back on the River Wensum. Ray was dropped off at the head of a private stretch of the river at 9.15 a.m. He decided that he would trot red maggots for chub. John then took Adrian three miles downriver, where they set up a ledger/pellet rig for chub again. At 10.10 a.m., Adrian had a bite and landed a five-pound chub that coughed up a ball of live red maggots when it lay in the net. Later, Ray confirmed that at around 9.40 a.m. he had hooked and almost immediately lost a good chub.

Just think about that. There were no other anglers on that beat of river, so those maggots were Ray's, for sure, and that chub had to be the one he had hooked, with no element of doubt. Talk about spooking the fish. It had flipped itself free and swum three miles in thirty minutes, only to peel off the current and get hooked again, this time on a different bait and a different method. Yes, it was hungry, for sure, but that doesn't alter the fact that it swam at six miles an hour to get away from a nasty situation. Six miles per hour! That would be good going for Michael Phelps, never mind a fish weighing five pounds and coming in at eighteen inches long.

Always respect wild fish, because they deserve it. Respect their senses and their survival skills, and you will catch more fish. But once you have caught a fish, why not move on, if you are able to? Let the rest of the shoal enjoy their day unmolested and they will be there to challenge you for sessions to come. Fishing is only natural history in action. That's its everlasting fascination.

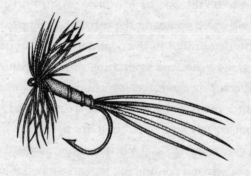

A hackled dry fly

CHAPTER SIX

WHAT'S IN OUR TACKLE SHED?

John: My tackle sheds have always been complete shambles, dumps of places where chaos has ruled. I've always tried to be neat, to wash bait boxes and have hooks arranged in a row, but all good intentions vanish like snowflakes in the sea and they have done since I was six. The bottom line is that I'm more interested in using gear than polishing it and admiring it. It's there to be used, and time is for fishing and nothing else.

Rats. Mice. They've been my horrors. Where there's bait and where there's shelter you get rodents, and my sheds have been inundated over the years. I even take them with me. Very many years back I caught TWENTY-SEVEN mice in my van over a single weekend. The traps were snapping every hour! I've walked along banks with mice hopping out of my pockets, and I've pulled on waders and been bitten on my toe.

BUT, the big question is what do I have in my tackle shed right now? And even more important, what do I ACTUALLY need? What I do know is that half the clobber I own has never been used and never will be. That's an expensive mistake to avoid from the off. As it stands right now, there are some seventy rods festering away and well over forty reels mouldering beside them. I've counted a dozen landing nets and scores of knackered bait boxes and chest waders I can't be arsed to throw away. That's a mess you'd do well to avoid.

Paul: *I'm in no way what they call a 'tackle tart', but there are items I have stored away that mean the world to me. Most of all, I have my dad's gear. I don't really use much of it; I just keep it treasured. It's all lying there safe for me to hand down when my time comes. Perhaps I'll give some of the collection to Cuddly Bob, as I like to call him. There's also a top of the range Thomas and Thomas rod Eric Clapton gave me. He even had COYS – Come On You Spurs – inscribed on the butt for me. A lovely thought, even though it hasn't translated into success on the pitch.*

I've got about thirty rods but only really use about ten of them; loads of reels that I've forgotten about; and fly boxes that haven't been opened for decades. In fact, I daren't look in most of the tackle bags I've got. There are assorted lures and flies for bonefish and tarpon that have rusted into oblivion. Obviously, I'm going to sort it all out ... tomorrow. Or the day after, actually. I'm very busy tomorrow.

My fly-tying kit is precious to me, and though I haven't tied a fly in anger for years, I sort of hope deep down I'll start again one day when time allows. I used to tie wet fly patterns for those sessions when I employed teams of three on an EIGHTEEN-foot leader, fishing them loch-style from boats on the big reservoirs. That's a reason I loved our Gone Fishing *episode on Lough Corrib. It took me right back, but I found it hard after a thirty-year lay-off. The wind was a bit tasty to start with, but it died a tad towards early afternoon. I was getting into some sort of rhythm when Director Rob got us into an island to film the lunch sequence. Filming and fishing don't always marry that happily, you know.*

I'm also very proud of my selection of Paul Cook grayling floats. Google him and see what he makes, because he's a real genius. I like fly but I like bait fished under a float just as much. In fact, I love to team the Paul Cook floats up with a titanium centre pin John gave me after series one of Gone Fishing *was filmed.*

I'd like to mention Jon Hall, one of the great chalk-stream river keepers I've known and admired for years. He was a bit sniffy when I first suggested a float on the River Itchen. But I had an epiphany (Del Boy from Fools and Horses *would say 'apothecary') some years ago when I realised that the chalk streams held wondrous coarse fish as well as trout. Now, Jon's well into roach and grayling himself. The method is a dream – watching that float investigate its way downriver, feeling the line peel off the old centre pin! I think I'd rather catch grayling on a float rather than fly for that moment when you strike into a big fish at distance and, crikey, it feels like a barbel or a grilse, even, on the end. It's traditional fishing: simple, beautiful, demanding and exciting as hell. What we love about it, to be honest.*

You go into a modern-day tackle shop and chances are it will be as huge as an aircraft hangar, and often it will be an impersonal experience. You'll feel like a spare part, wandering around, feeling gormless next to the self-styled experts lounging around the till swigging coffee and laughing with the easy confidence of anglers who catch leviathans with every cast. We feel intimidated, so chances are you do, too, especially if you are starting out or coming back in after a career/family break. How do you even start to ask for advice when the whole building groans with gear that you have no idea how to use? It's like these places are designed to befuddle and bemuse.

If you can, find a nice, happy, helpful, low-key tackle shop – one that's a bit old-style (they do still exist, honest). In Herefordshire, there's a belting game outlet and at least three coarse dens of delight where down-to-earth people will listen to your needs and sell you what you want. But wherever you shop, what do you want if you are starting out?

Consider three questions. What types of water will you be fishing most? What species of fish will you spend most time pursuing? And what methods are you likely to want to use in the catching of them? Do your research. Talk to anglers on the bank where you are likely to start. If you can, go to tackle shows and country fairs, where you will find help.

It is probable that you'll start with coarse fish on local lakes, pools, pits and rivers, so let's have a look there for starters. If you are not going to fly fish, then we'd suggest you look at an Avon-type rod, which will do you for most fresh-water fishing. It will be around eleven feet long and have a test curve of around one pound (that just means it takes a one pound weight to bend it). Team the rod up with a decent fixed-spool reel, about 3000 or 4000 size (don't worry, the tackle shop will know). Buy two spools, and load one with four-pound line

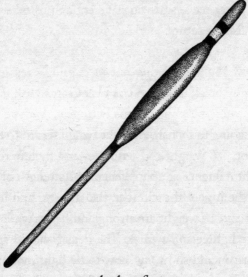

An Avon float

and the other with ten-pound line (ask the tackle dealer to do this for you).

Add floats, shot, leads, hooks and forceps to your bill, and you are ready to finally think about bait (see the list of terms in Part Three for more information). But that, of course, depends on what you want to fish for. With gear like this you can fish for decent-sized carp and tench in lakes, chub and barbel on rivers, and roach and perch everywhere. You can even cast lures for perch and smaller pike. This one rod with reel and the differing lines and baits will take you a very long way, certainly far enough to know if fishing is for you. Even if you want to take the sport far more seriously, there's no need to go overboard. Keep the Avon, add a thirteen-foot float rod, two light carp rods (which will do for pike as well) and a couple of extra reels, and you have enough to take you well into your fishing future.

FLY FISHING

If you are going to fly fish, and why would you not, start with a nine-foot, 5/6 weight rod with reel and fly line to match. The weight reference simply refers to the strength of the rod, in which the higher the number, the stronger and heavier it is. So, 8, 9 and 10 weight are strong and heavy, while 2 and 3 weight are lighter and weaker. The dealer should spool the reel up for you and show you how to tie light monofilament line to the heavier fly line. Ask for advice on flies to buy, and you are almost ready to go.

It's a great idea to get casting lessons to begin with to learn the basic; lots of day ticket fisheries offer this. Venues will also help with tackle and might even sell it at a reasonable cost, too. Once again, as you progress you might want to add lighter or heavier rods and reels, or even salmon fishing gear, but by then you will have experience, confidence and contacts, and your fishing career will be launched.

CLOTHING

Fishing can be a miserable game if you are wet and cold, so clothing is vital. John will never forget the winter of 1963, fishing through arctic conditions in shorts, flimsy wellingtons and an anorak that soaked up sleet like blotting paper. Today, of course, there is specialised clothing for cold, mild and hot weather.

Pay attention to terms like 'layer systems'. Buy good quality and you'll learn to laugh at whatever the elements throw at you. Breathable waders are great, too, although a bit of an investment.

John: I do have one serious thing to add here, however. I'm on a one-man campaign to outlaw wading boots with felt soles, which have been de rigueur in the salmon fishing community forever. I guide, in part, for a living, and a vast number of the potentially serious injuries I have witnessed are because of this material. It's a killer, literally. Felt might be all right on sand and gravel, but it's no better than cleated rubber. And on mud or wet grass, felt offers no grip whatsoever, even with studs hammered in. It makes getting in and out of most river situations all but lethal. Felt soles are typical of those old salmon shibboleths that have been alive, kicking and unquestioned for decades.

Paul: I'd agree with that. I consigned my felt-soled wading boots to the dustbin of eternity a long time ago. There are indeed much better composite-type soles now that grip the river bed and wet, grassy banks equally well. But I would add that wading can be a treacherous process, so tread carefully, literally!

And finally, never neglect headwear. Even Bob's trusty pork pie hat fulfils many functions. It keeps the rain, sun, wind and even the snow off him. It keeps flying hooks out of his skull and its brim gives protection from direct sunlight and glare. We'd only add that a completely waterproof hood is perfect for downpours and a thermally insulated beanie can make light of the worst of the winter.

In short, try to buy the best you can afford. Insure it. Look after it, and it will look after you.

CHAPTER SEVEN

OUR BAITS

John: I mentioned the eternal battle I have with mice in my tackle sheds. Well, a dead mouse is a bait hard to beat for the biggest chub that swims in your river, but I understand your reluctance to try this one.

Unsurprisingly, you can catch fish on a myriad of baits. I've seen the weirdest lures flung out all around the world, but never anything to compete with Batsok's Special in north-west Mongolia back in 1994. A gang of Czechs and I took a flight in a single-propeller Antonov plane from Ulan Bator to the deserted steppes of Khovsgol Aimag province, capital city Moron (Paul: Home from home for John, then) – a journey of two nerve-shredding days. Our pilot, our own Red Baron, navigated 10,000-foot-plus peaks and hundreds of miles of empty desert to land us on the Shishged River flood plain, by a huddle of white tents – the gers of a family of Mongolian nomadic herders.

There was wonder, confusion and general amazement as we stepped from the plane. The children hid, and only the camp leader, the fearsome Batsok, strode forward to meet us. The Red

Baron explained that we had travelled across Asia to find and catch the mighty taimen, the great landlocked salmon of the Far East. Batsok listened and his face creased into a smile. 'Sharks here,' he said, pointing to the River Shishged, which boiled and foamed in fury across the mountains into Siberia.

His little daughter, as instructed, then shyly approached with a battered teddy bear and a length of rope attached. Like an Asian Rowdy Yates from Rawhide, *Batsok whirled the bear round and round, and flung it out into the current. Hand over hand, he pulled back the rope and the bear breasted the cauldron of water, pursued by a bow wave that engulfed it in an eruption of foam. A huge taimen broke the rope to a shriek of horror from the tiny girl. Batsok only laughed, beamed some more and said again, 'Sharks here.'*

Paul: *Both John and I are old enough to remember when reading books was the way to enlightenment and when people had enduring heroes. That's one reason why we both go back to old angling tomes like* The Compleat Angler *or* Mr Crabtree *and all the great works of the past.*

Another reason is that it's comforting to think that our sport has been enjoyed more than any other known to history: anglers have been at it for millennia, sometimes for food, sometimes for sport, sometimes for both. You can't help but wonder how good those fishing ancients might have been, and what skills and knowledge have been lost from a time when tackle was rudimentary and sheer ability was needed to compensate. (Come to think about it, 'sheer ability' isn't a description you see often on the riverbank these days, is it?) Baits? Boilies and pellets have only been around a few decades, and sweetcorn only just before that. Would anglers five hundred years ago have used bread or would they have needed it too desperately for food?

Two baits from the annals of angling's past have always fascinated me. Bullock's pith, for starters ... No, not a typo, honest. Robert Venables (no relation to the great Bernard that we're aware of) wrote about it in 1622, saying something like it's the marrow in the bone of the ox back. You take it out carefully 'and be very tender in taking off the tough outward skins but be sure you leave the inward white skins safe or your labour is lost'. Health and safety probably rules pith out these days, but it worked wonders on chub in the days of the first King Charles. Tallow is another forgotten favourite. According to ye Angling Times *of 1832, or whatever Smith's sold then, this is the hard substance made from animal fat that was used to make candles and soap. At least if tallow caught you nowt, you could always lick it.*

If we look back over the history of angling, there can be no doubt that the most successful baits have been naturals, the foodstuffs that fish come across in their everyday lives. Some baits of this sort are only attractive for short periods. Perch, for example, will gorge on tadpoles in the spring, while all river species relish mayflies in season. Salmon often swallow lobworms deep into their throat, which makes unhooking a difficult task. For that reason, worms are barely ever used today, when we all aim to put salmon back unharmed.

The moral is to keep alive to the infinite possibilities nature offers you, and you'll profit from baits that are free, easily attainable and desirable. But you have to be aware of the forces of change. Our hero Izaak Walton and his friends would have used both swan mussels and native English crayfish, the latter for chub especially, but now both are too threatened to be considered for hook service. They would have used other kinds of bait, too, including woodlice and caddis grubs. Turn over stones or wooden debris in the margins of a river and there the caddis live in little cocoons made of sand, twig and gravel. The caddis is the larva of the caddis fly; prise out the white or green grub inside, stick two of them on a size 14 hook and you'll catch everything from dace to trout to barbel.

Another old timer of a bait is silkweed, the soft cotton-wool-like stuff that grows in quicker water come summer. Victorian anglers used it under a float, drifting down the current. Fine, though it's a bugger to keep on the hook and we've never actually seen a fish caught on it. (*John: Hmm. Not quite true. I saw a single comizo barbel take silkweed on the Guadiana River in Spain back in 2001. Fluke? I think not.*)

LOBWORMS

Excellent as these baits are (sometimes!), the aforementioned lobworms are what we really love. Not red worms or brandlings, but big, fresh, juice-rich lobs, perfect for every coarse fish that swims. During a flood, thousands of lobs are washed into any river system, and every single type of fish adores them and recognises them as a serious treat. Two lobs, hooked near their head on a size 4 or 6, are just about the perfect mouthful for perch, chub, barbel, carp, bream and even big roach. Note that we're not keen wormers for trout and never for salmon these days when numbers have fallen so severely.

Chopped lobs are a powerful attractant if you have enough to spare (we know they are expensive these days and not everyone has a garden to dig them from). In the days of yore, both of us went snitching worms from parks and cricket greens at night, when dew falls thick. The lobs lie in their thousands, luxuriating in the moist air, and the trick is to catch them near the tail before they retreat to their underground tunnels. This is a sport in itself, but does anyone think to roam the urban grasslands of modern-day Britain, torch in hand, suspicious cloth bag up their jumper? (*Paul: We'd seriously advise not to.*)

SHOP-BOUGHT BAIT

Bread is best. Nothing beats it long-term for roach, chub, carp, tench and even barbel and trout. It is cheap (ish), even in these inflation-crazed days, and easily obtainable and very visible. Bread flake is fantastic. Simply pull a piece of bread about an inch square from a slice of Mother's Pride or the like. Carefully but firmly press the bread around the shank of the hook. Try to ensure that part of the bread is not squeezed but remains nice and fluffy, and always keep the point of the hook exposed. With practice it will stay on for twenty minutes and grow more attractive as it expands and gets all sloppy.

Yum. You can make a paste out of wet bread and mix it with cheese. You can also use the crust of an unsliced loaf as a floating bait for carp and chub, and all these variations have their moments. Use your loaf and you won't go far wrong.

Supermarkets can provide you with a lot of options: sweetcorn is brilliant for tench, carp, roach, bream and barbel; luncheon meat is a fine bait, especially in swollen rivers. Cheese is fine, too, but choose rubbery stuff that stays on the hook – those cheese slices that you'd never consider eating after the age of six have worked for us.

MAGGOTS

One option food shops do not offer these days is maggots, the larvae of bluebottle flies. John's Victorian grandmother told tales of buying maggots from the butchers – and John tried to do the same when he was a young fisherman starting out, but to no avail. There's a surprise!

Wherever you buy them, maggots really are truly magnificent bait. Along with bread and worms, maggots would probably do us for life. Red maggots for tench; white maggots in coloured winter rivers for roach. Take care of your maggots and don't let them get hot in summer or wet ever – rancid maggots that crawl everywhere are never to be encouraged. Keep them in a fridge, or at the very least find somewhere to keep them cool.

Maggots are natural baits but you buy them commercially. They also look much like caddis grubs, creatures all fish adore.

They are also what we 'pros' call a particle bait, which means there are lots of them, and this profusion drives fish wild with glee. On *Gone Fishing* business, neither of us leave home without maggots, which catch everything that swims. Learn to bait hard with maggots: you cannot put too many free offerings into 90 per cent of waters. If you can feed a river swim for ten minutes without actually fishing it, chances are you'll get a bite first cast. But remember: Don't Spook the Fish! Controlled patience is what you need – a mindset that allows you to relax then burst into action when the time is right.

PELLETS AND BOILIES

Pellets and boilies are the super baits that dominate coarse fishing, and every shop sells a bewildering amount of them. Pellets burst on to the scene in the nineties as a great bait for barbel and chub, because they are convenient to use and the fish oils in them are madly desirable, it seems, to the fish. Boilies date a little earlier, and in the eighties we largely made our own from recipes that included eggs, semolina, 'tutti frutti' flavourings and neon orange dyes. These ingredients had to be mixed, rolled into marble-like balls and then boiled to give them a hard skin. (Boiled baits – boilies! Get it?) All serious carp fishers' kitchens stank the street out, so it's a relief today that tackle shops offer a mind-boggling mountain of different flavours, sizes and colours.

To clarify, both baits are generally fished off the hook, attached to hairs (see Part Three: The Nuts and Bolts of What

We Do). Most hook baits will be between 8 and 18 ml in size, depending on the fish being pursued, though 2, 4 and 6 ml pellets are useful for smaller fish and for loose feed thrown out as an attractant. So-called halibut-flavoured pellets are our favourite, but where to begin with boilies? John fishes with them more than Paul, and he's happy with 12 and 15 ml for hook baits, generally fishy flavoured, like crab or krill, and coloured dark red or brown. That's a gross simplification, we know, but if it's carp, barbel, chub or tench you are after, we promise you won't go far wrong.

A final thought about bait: you never know what fish will fancy or when. Take slugs. We've caught flip-all on them, and yet pals of ours have had magic catches with the giant black ones that scoff a lettuce each most nights. In short, it's a mistake to write almost anything off.

As a kid, John was a junior member of a Stockport fishing club, and he and his mates went on all the Sunday matches. The Bad Lads sat at the back of the coach, watching the crates of pale ale being loaded into the boot, and then they'd be off to the Trent or the Witham or to any exotic venue away from the dark, satanic mills of the north.

Weigh-ins at the end of the match were the best fun: the lads would follow the scales men around, oohing and aahing at anything outsized. George, the oldest angler, was invariably asleep late in the afternoon and would tell the assembled company that no, he hadn't caught a sausage. After two seasons of this, on the Trent, the Bad Lads crept up to George, asleep as usual at his peg, reeled in his line and impaled a whole pork sausage on the hook. It was an absolute python of a thing, and they lobbed it back into the flow and slipped off, consumed by silent, eye-watering, shoulder-shaking mirth. An hour later, they found the aged angler awake, pacing the bank, eager to weigh-in his match-winning chub. Blooming great thing, said George. And in its mouth? A whopping big sausage!

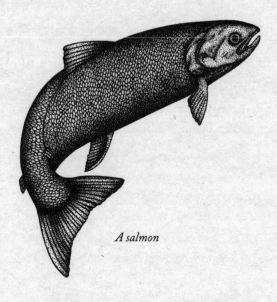

A salmon

CHAPTER EIGHT

OUR FLIES

John: Late-summer 1964 changed my fishing life. Throughout August one parent or another would drop me at the village of Tintwistle, on the fringes of the Peak District, and from there I'd walk to the string of reservoirs where I had discovered that exotic species the brown trout. Most of the first two weeks I was happy drowning worms on the lowest of the waters, quaintly named Bottoms, but the days were long and often dull, so increasingly I looked to the hills and the next reservoir in the chain, the dauntingly beautiful Valehouse.

This was a real step up for me geographically – a bloody long walk, that is – and as Valehouse was fly only, piscatorially too. It was one of life's lucky chances that a distant uncle had awarded me an ancient fly rod, reel, half useless line and the remnants of his fly box the previous Christmas, so I was miraculously equipped for this new challenge. I barely remember seeing a soul fishing during that time, though the water keeper and I became close and he passed on the rudiments of this new side of the sport to me. I caught a couple of fish on a Red Tag, a very traditional fly pattern, and this gave me some sort of faith in creations of fur and feather, though a part of me still hankered after a lusty lobworm.

Each afternoon, quite late, swarms of wobbly daddy long legs – crane flies, to the experts – would find themselves blown off the surrounding moorland and into the water, where they struggled, drowned and were picked off by trout. This daily occurrence encouraged me to seek out the windward bank around 4 p.m. and wait for the breeze to blow the armada of daddies inshore. Carnage! The trout by now knew the score and arrived in troupes to mop up the floundering insects. I found that an imitation worked exceedingly well if I put as long a cast as I could manage out parallel to the bank … and waited. There I'd sit, on the stones, in a tremble of excitement, waiting for a trout's neb to break surface, for the daddy to disappear and my line draw tight.

To me then, this was a dazzling approach, far more thrilling than my worming could ever be. The fact that these wild, wonderful fish could fall for some bristle concoction from a dusty old box only increased my exhilaration. It was, we would say now, a light-bulb moment – an explosion of insights went pop in my young head. But before I could blink and take in all this excitement, it was the first week in September and the last of my daddy days before the loathed autumn term would begin once more …

The evenings were cooler now and the daddy hatch less prolific, but still the trout queued for their supper, which I was delighted to serve them. But this night, I see it even now, the trout's neb was enormous, the rod bent like never before and the reel screeched into a cacophony of grudging life. I played that fish like my life depended on it, even as the light faded and I knew Father would be waiting, furious at my delay. But I could not let go; this was to be my moment, the crowning glory of the summer. Inevitably, the trout had other ideas, and at the end of a very long line, the chafed nylon parted and my heart broke along with it. That crushing disappointment was the end of my Valehouse era, but a letter came for me towards the end of the month. The keeper wrote to tell me that a trout just shy of seven pounds had been washed up by the dam. Enclosed in the envelope was what remained of my lost fly, two long daddy legs still intact. Sixty years on, that loss hurts to this day.

Paul: Once again, I was lucky having my dad to mentor me through my formative fishing days, and there really is no better way to learn the arts than through a parent's guidance. He'd been all about bait until he saw a bloke with a fly rod when he was young, and he was so intrigued he started to read up about it until it became his passion. But he still enjoyed taking me bait fishing, so he took us to the Usk and plonked me on a slack with a worm on the end while he went off, fly rod in hand. That's where I landed my first ever trout, and Dad came back with a couple more, which we ate fried with butter.

I don't suppose any of us would dream of doing that now, but I wonder why not? Catch and release is the way everyone goes now with wild fish, but that certainly wasn't the case forty, even thirty years ago — and in the centuries before this one anglers ate almost everything they caught. Perhaps that's our best measure of conservation initiatives — when we're able to once again eat what we catch if we choose to. If anglers could kill a fish each time because there were so many fish, wouldn't that prove that all our efforts have eventually succeeded, and all our catch-and-release restrictions have paid off? Imagine an abundance of wild salmon again, with big numbers surging up all our rivers once more. This might even mean an end to the dire salmon farming operations that are causing so much environmental carnage. You can't call us controversial in any way, can you?

Anyway, after that outing, Dad actually made me a fly rod of my own. He fashioned the corks, whipped the rings, everything, and that present was right up there with my first leather football and a Scalextric set. He didn't really teach me to cast, though, and I never got instruction until I began to salmon fish and the great god Hugh Falkus taught me the Spey variations. That's why my single-handed casting can look a bit ropey at times.

I actually lost that rod on Lough Conn in Ireland a while later. It was a scarily rough day out there and my dad and I were panicking. My line got caught in the propeller somehow. It whipped out like I'd hooked a salmon, and in an instant rod, reel, the lot disappeared over the side. By then I was really into fly fishing and loved the fact that I could have a rod and tackle bag all ready for a quick trip at a moment's notice. This was the period of the boom in stillwater trout fisheries, and I even took my old coarse-fishing mates with me. I got Martin Bentley to come fly fishing with me one day. He was standing behind me while I showed him the ropes, and we got a lure in his right ear. We pulled that out, he had a crack and got the same lure in his left ear. He went home with two bloodied earlobes like he was wearing earrings. Not sure if he ever went again, to be honest.

John's fly selections are basic but tried and trusted, and we hope emphasise our belief in keeping fishing simple.

There was a time when he was slightly in awe of the fisher with a boot full of fly boxes, but that was many moons ago and now he's happy to go out with flies in little more than a matchbox. He's got a handful of mayfly imitations, a couple of dozen red and black buzzers, a few Muddler Minnow-type lures from the 1980s and perhaps fifty salmon flies from the years he had the obsession (and the money) to fish for them. Today, though, what fly fishing John does is rather restricted to natural-born brown trout a few times a year, barbel a good few times in the summer, and grayling pretty much most of the autumn and winter. As a result of this fixation on wild river fish, often in colder weather, he has come to rely heavily on nymph patterns fished subsurface.

This began way back in 1996, when John took a group to the Czech Republic on an exploratory barbel quest. For several days the team caught nothing on any of the usual baits, and then one morning their guide, Czech international angler Franta Hanak, turned over some rocks in the margins, picked out some caddis grubs and put them on a size 14 hook. Bingo. Barbel after barbel. It was another lightbulb moment, but better was to come.

Towards the end of the trip, John was fishing a weir with a light lead and caddis on the hook. There were fish there, judging by line bites and quick pick-ups, but three hours had gone by without a strike in earnest. Franta wandered behind him and asked if he could possibly try with his team of three nymphs. John was glad of the break. Franta made six casts and six times his strike indicator dipped. One trout. One grayling. One bream. Two barbel hooked and landed, the third and last barbel breaking free. 'Hmph,' said Franta. 'Too many coarse fish. I go downstream.' A lesson learned indeed: never before had John taken seriously the fact that an artificial fly could catch as many or more fish than a juicy natural bait. Sometimes what is on your hook doesn't matter as much as how you present it.

For John it was nymphs all the way from that day forward for the fish he loves to catch. He's got a few hundred flies, many from days spent in Eastern Europe, but most have no name and are simply generic, looking like any unspecified invertebrate a fish could come across while wandering the bottom or adrift in the flow. Most are coloured brown or black or dark olive. Some have a flash of red. They are tied on hook sizes from 8 to 16, generally, though a few heavy boys for deep-lying barbel lurk on even bigger hooks than that. With this motley crew he is happy to take on any fish in any type of river he is likely to find. All his attention goes into reading the water, locating his targets, presenting his nymph at the right depth and at the right speed, and then seeing the take. These considerations consume his energies entirely and let others engage more fully with all the flies the game world has dreamed up. Which is where Paul comes in.

Paul: If John, or anyone else, come to that, expects chapter and verse on fly choice, forget it!

Looking at trout first, I'm completely happy to fish with nothing other than a Goldhead Hare's Ear. The pattern is so much like a natural invertebrate that I've never found a wild or stocked trout that won't take it with complete confidence. The one thing I would emphasise is that size can be crucial and to take a really good range – a big one can even double up as an effective lure.

I'll add black or dark olive Klinkhammers to the list, too. These have revolutionised dry fly fishing – you can see the top bit easily and the body of the fly lies deep in the surface film, looking for all the world like a hatching caddis larva. It's a great fly to fish on a slow day when nothing is showing, because trout will come up for it out of nowhere. And, of course, you can fish it klink and dink – that is, with a Hare's Ear attached beneath it, set at any depth from the bed up. Some call this New Zealand-style, but all it means is that you have a visual aid that screams out when the nymph is taken, and there's always a big chance that the Klinkhammer itself will be snaffled off the top. Two top flies tied on different hook sizes and that's all you need. It's all I need, anyway.

I'm as basic when it comes to salmon flies, too. Having said that, it's always worth ringing the changes. If a salmon has seen countless two-inch Willie Gunn flies, it might make sense to change it, if only to give the angler renewed confidence – and confidence is key. So, no, you can't restrict me to one or two flies, that would be a ludicrous idea, but for much of my salmon sessions I'm totally happy with gold-bodied Willie Gunn patterns with either a black or red cone head. Just like my Hare's Ears, it's having Willie Gunns in different sizes and weights that's the critical thing. It's such a versatile fly, and I'm dead confident with it on the clear waters of the Dee or the darker waters of the Thurso. I can fish tiny patterns in high summer on really skinny rivers, but I can go to three inches or more on a sinking line when levels are high. The only proviso I'd make is that I'll always have a Sunray Shadow ready to go on if more conventional approaches don't work for me. If I've worked a pool hard with a Willie Gunn, I'll often try a Sunray fished down and across with some speed. It can work wonders, livening a whole pool up. Salmon will slash at it, but they don't always take. While for me it's one of my cardio workouts. I'll be moving quickly, stripping back fast, and my heart is always in my mouth.

So, that's four flies that would do me for pretty much everything on a game river. I'd add the Medicine fly, too, concocted by my guru Hugh Falkus, for sea trout. But, as you can see, I put my faith in presentation more than pattern.

CHAPTER NINE

OUR LURES – WITH GUEST ROBBIE NORTHMAN

John: Between 1990 and 2010, on my travels through Europe, I realised that anglers over there were so far ahead of me with fishing lures that I was a lifetime behind and would never truly catch up. I reckoned it was like skiing: if you didn't start as a kid, you might as well forget it.

Two big, eye-popping sessions told me all I needed to know about my ignorance. In 2002, I was in Sweden fishing the coastal Baltic for pike with three Scandinavian guys. In three days, on lure, I had 4 fish. Between them, they had 210. Then, in Eastern Mongolia in 1998, I was fishing for Amur pike with Czechs and Slovaks, and again it was all about lures – and again I was slaughtered. This I still find hard to believe. One of the oldest and hoariest of my companions, generally drunk on vodka or slivovitz or both, fished standard silver spoons but in a way I'd never considered before. Whereas I – and probably you – would cast, let the spoon sink a little and then retrieve it, my old Slovak friend did nothing of the sort. He cast and let the four-inch spoon sink to the bottom and stay there for two minutes at least. Only then did he lift the tip and crank a couple of times before letting the lure settle for a minute or two again. This way, each cast/retrieve lasted upwards of fifteen minutes, but get this: while I caught nothing, in two hours my man had three Amur pike and lost a huge creature that neither of us ever saw. If that's not enough, ALL the takes came when the spoon was lying inert on the bed of the slow-moving river. Yes, those pike looked at and picked up immobile bits of metal that had no life in them what-soever. Unbelievable. What do I know? Certainly not enough to tackle a subject like this. Let's see what Paul has to say on the matter, but chances are we might have to call in an expert ...

Paul: *Yes, John, I'm no expert on lure techniques, but I've enjoyed it hugely in my past when I used to do a great deal of spinning. But I have to get this joke in: these days I am more spinned against than spinning, but I have the happiest memories of what is a great skill if done right.*

We're going way back to a time when my dad and I fished the Welsh rivers for salmon and sea trout with worm and spinner quite happily. We didn't have access to the top beats, nowhere near, and it was a case then of staying in a caravan shoved in some farmer's field. We had no ghillie, no real knowledge, and a week fishing for migratory fish that might not have been in the river anyway; it was a real commitment. We'd often go days without a pull. After all, even if migratory fish are in the river, that doesn't mean they'll be feeding, and generally conditions would be all wrong as well. It was a big struggle for us, but we expected little else, to be fair. We knew we had to work at it and the results would come. And they did.

This little story is all about my first salmon on a lure (a Toby). I call it my Teifi Tiger Toby Tale (how about that for alliteration). It happened one year when we found that lovely river just right at long last. It was fining down after a long period of high water, and I got to a great-looking pool, made two casts and, Bang! It looked a huge fish for a small river, and even now I can smell that damp, cloying scent of the early morning woodland. According to family legend, I was so excited that when my dad arrived on the scene I literally jumped into his arms. I've no recollection of that, but I know it'll be true – I just went into some sort of mental meltdown, brain turned to mush. What I do know is that in those days my fly fishing skills would in no way have caught me that particular salmon The fish was lying too far off and in water too deep for me to reach it, but the Toby's extra weight did the job. And thank God it did. Now I'll hand over to Robbie. He's the lure maestro, John, not us.

Robbie Northman is an extraordinary angler for any age, talented beyond compare and dedicated to the level of obsession and back again. When we fish with him we are agog; he just seems to catch fish wherever and whatever the weather and conditions. His focus is unwavering. His self-belief is unshakable. His angling imagination is without limits. His range of skills is immense, his patience endless and his intuition and instinct are proved correct over and over again.

We've both fished with Robbie on multiple occasions, especially John, and we have barely known him ever to fail. Robbie proves time and time again that fishing talent is absolutely real, that luck and opportunity only play their part to a certain point, and then it is sheer ability that takes over. It's all-action fishing that Robbie takes to another level. He's about fly fishing, stalking, sight fishing, surface fishing and, above all, lure fishing, where he is a true artist. He can work a beat of river like no other angler we have even seen, which is amazing in a lad so young. How did he get to be so good so soon? It's almost like he's our hero Izaak reincarnated.

The concept of using fish imitations made of metal, wood or even bone to attract and hook predators goes back millennia. Pike especially attack lures that resemble food, mostly fish but also frogs, rodents, birds, and even snakes and lizards. The array of lures on the market today is vast. The American, Scandinavian and Eastern European anglers in particular have taken lure design to dizzying heights, and the choices facing us are baffling. An expert is needed as a guide through the labyrinth. Over to you, Robbie.

Robbie: It's great to be asked to contribute this intro to lure fishing because I find it frustrating, absorbing and thrilling. I simply love it! It's also the best and fastest way to locate and catch predatory fish. It's all action and full of thrills and spills, perfect for modern lifestyles. Here are my ten top lures and observations.

1. *Soft plastic lures/fish patterns. There's an absolutely endless range of these, from one centimetre to forty, and models like pintails, fluke designs, pulse and paddle tails. Try to match the size to the quarry and think about how the lure will work in the water column. Tailor your retrieve style and speed and the running depth to how you read the situation.*

You can jig soft lures absurdly slowly – let them hit bottom, pause, retrieve and wait for a take. You can retrieve normally; that is, horizontally when predators are active, often striking on the surface. It's a method without limits.

2. ***Soft plastic lures/creatures.*** *By that I mean crayfish, worms, crabs, shrimps, whatever the imagination conceives and can be manufactured. Many are great for saltwater work, and worm and crayfish patterns are tops for perch and zander. Work them with low, slow, deep retrieves. Try drags and gentle pulls and intermix with aggressive hops and bounces. Visualise the movement you are generating. Think about and imitate the panic responses prey give out when stalked. Every cast should be an adventure.*

3. ***Soft plastics buoyant.*** *Both fish and creature plastics can be bought as floating lures, and these are perfect for perch and for the US and European bass market in their smaller sizes. A slow drag lays a buoyant lure horizontal, but if you pause the retrieve it will stand up and look tantalising. Hop them. Bounce them. Quicken them up to grab attention. Then a long, static pause will seal the deal.*

4. ***Spinners, spoons and metals.*** *They're simply timeless. Spinners and spoons have lasted through the ages and are every bit as effective to this day. They are all worked well with a straight retrieve, but mix it up with pauses which allow a flutter. Great in sunlight, good in deep water and along the coast.*

5. ***Crankbaits.*** *These are hard lures with big action and depth controlled by a lip/diving vain on the front. They come in different sizes and colours and either in sinking or floating*

patterns. They are great for cast and search sessions, when you are prospecting waters and on a recce. I like the floating models. You can dive them to set depths or find the bottom, and then as you pause they'll rise up through the water. You're investigating the whole area with a killing zig-zag motion. Sinking cranks are great for extreme depth or fishing close to structure. Suspended cranks have neutral buoyancy and hover in specific attack zones.

6. ***Jerkbaits.*** *These are again hard lures and you impart your own distinctive action on them. Manipulate them to achieve wobbles, wide glides and erratic darts. There are loads of varieties, but imparting big action is always the game here. Timing the retrieve with a figure of eight tip action creates a curve. Swooping the rod tip at alternating angles will ensure the crank comes back in a wide curve. Fast whacks on the rod tip give an erratic, irregular movement which predators salivate over.*

7. ***Topwater treats.*** *Perhaps the most exciting lures out there, if not always the most effective. Try wild, wacky animal patterns like rats or ducklings aimed for pike and something more subtle for the rest. There are simple pencil-style bodies that you have to work like you do jerkbaits. There are popping models that make a real noise when zapped back. There are prop models that churn and gurgle on the retrieve. Crawlers feature wings that make the lure swim and rock from side to side.*

8. ***Colour.*** *Our lures come in a vast range of shades, but I always think natural is a good place to start. You rarely go wrong with straightforward prey imitation. Try both*

bright and dark lures at night and when the water is coloured. Take black and white as the two extremes, with blue, pink, firetiger, green, orange and brown in between. To summarise, I fish natural, blue, green and firetiger colours in clear water and hotter colours when it is murky. Black and white work in both, and I remember John telling me about sessions for ferox trout in crystal-clear Scottish lochs that would only have black lures in bright sunlight.

9. *Sizes. You always have surprises and there is never an unbroken rule here, but by and large: for perch I put on lures between 2 and 9 centimetres; I'll go 7 to 15 centimetres for zander; 10 to 20 centimetres for sea bass; and anything between 10 and a mighty 40 centimetres for pike. Some huge Low Country pike are coming out on lures even bigger than that if you have a couple of hundred euros to spare.*

10. *Work. That's the key. But hey, it's nice work! Use your imagination and work your watercraft skills. Keep trying, and remember what you have used and how. Trial and error is what makes the top lure angler in the end. Listen to advice. Watch others. The learning curve is steep and endless. I live to lure and I love every minute.*

CHAPTER TEN

LUCK

Paul: It pays to be lucky and some people seem to have it on tap. Some guys have all the luck, as Robert Palmer sang and Maxi Priest echoed. Luck usually only gets you so far, though. You have to do the legwork in most walks of life before you can hope for luck to come your way. A top striker might nick a lucky goal off their shin in the six-yard box, but they've dedicated years of slog on the training pitch to get to that point. Luck is not something to rely on. Which is why gambling is, let's face it, a mug's game.

Where fishing is concerned, though, a complete novice can, on the odd occasion, find themselves being very lucky indeed. Though once again with the caveat that somebody needs to have helped them set their gear up. Once baits or lures are being fished, though, there is a chance that a total beginner might find themselves attached to a big fish. Whether or not they'll land it without some expert help is another matter. Obviously the beginner can't hope to compete against an experienced angler. There are so many variables with species, venues and tactics that as you begin to think more about it the poor novice hasn't got a cat in hell's chance. (While we're here, by the way, why is the cat in hell? Animals don't have souls. Apparently. And that raises another question: how (the hell) does anyone know this? But I digress. Back to luck.)

There are two types of luck, of course, and we've all been on the end of both good and bad luck on the bank and elsewhere. My most fortunate catch was a salmon I became entangled with on the River Dee. It was a cool spring day and I waded slowly down the river, a relative novice to salmon fishing. The lambs were gambolling in the field behind me and the daffodils were heralding the spring, as they do. Well, they might have been – I didn't look round once. I was entirely focused on my fishing, especially as I'd seen a fish move on the far bank. I carefully covered the spot but became aware of a dragging sensation around my feet, which I thought must be an old tree branch trundling down the river bed. I trudged out to inspect what was happening and saw that I had become entangled in some yellow spinning line. I handlined it in to find I was, astonishingly, attached to a salmon! Admittedly a kelt, which is a spawned and recovering salmon and has to be returned by law. Which it duly was. This salmon fishing is easy, I thought. I don't know what all the fuss is about!

These days almost all salmon go back in UK waters, which brings me to my first example of bad luck – on the same river and with the same species. I was fishing at Park, a beautiful beat on the Dee, and I had been hoping to buy myself a new fourteen-foot double-handed fly rod. An expensive new one made from graphene and 'space-age' carbon technology was just about coming onto the market (yes, I'm a sucker for this stuff, too – tackle catches as many anglers as it does fish!). Luckily, I knew the manufacturer, a well-known casting expert and instructor on the Ness, Scott Mackenzie. I'd used one of Scott's rods for years and loved it, so was keen to try the new version. He kindly said I could borrow a demonstrator (one of the first run) and sent it down to me from his scary part of Scotland.

I tackled up with serious anticipation, not least because the rota had me on 'The Kirks', a lovely, fishy beat at Park, which always holds salmon, and some big specimens, too. As I approached, I saw there was a bloke fishing the opposite bank, but his water ran out just as mine started, which was just as well because a good fish moved out in midstream. I got in just above the rise so I could get an initial feel of the rod before I covered the fish. Ping. Out went the line nicely. A few steps downstream, a few casts and, Wham! The line tightened, and I felt the magical take of a salmon on a fly. Not much can beat it, including a winning goal, off the shin or not. Meanwhile, the guy opposite must have been cursing me silently, unless he was a better human than I am (this is, of course, very likely, but his moment would come).

I let out a smug, 'Yes,' and played the fish very well, the rod doing its work perfectly. There was a convenient little sandy bay with shallow water where I could effectively unhook the fish without taking it from the water, and after a hard battle I was able to guide it there feeling very pleased with myself. Hubris is bad, we are told, and pride comes before a fall. What happened next was a comedy of errors and I hope opposite-bank man enjoyed every second. I would have.

I gently drew the fish into the bay but was so pleased with myself that I failed to notice the trajectory of the rod tip, and before I knew what was happening the (very expensive) top section (of 'not my rod') snapped as I drove it into the rock I hadn't noticed on the bank! 'Oh,' I panicked. 'Now what?' The fish sensed my indecision and decided to head back to the North Sea through my legs, with great aplomb. Nutmegged by a fish.

My fellow brother of the angle opposite must have been in stitches and thanking whatever god he believed in for my come-uppance. I finished my performance by trying to grab the line, and the salmon shot off from whence it came, leaving me with no salmon, a broken rod that belonged to somebody else, and a healthy slice of humble pie for my next meal. I tackled down with as much dignity as I could muster (none) and got out of there sharpish, all while contemplating how I would explain it to Scott. At least I bought a couple of his rods (I'm still recovering from that, too).

One other example of the twists and turns of outrageous fortune came in the form of a moral lesson when I was a teen-ager. I had bunked off school, played truant, whatever you call it (we actually called it 'playing the hop' round our way), but I didn't fancy hiding at home all day. So, I hopped on a train to Cheshunt to fish the wonderfully named Friday Lake.

Friday Lake was (still is?) a difficult venue for us youngsters, but it held big fish. My mate Daryl had landed a pike of 29 pounds after a Herculean four blank sessions, and we knew there were some big carp in there, too. So, armed with a nicked loaf of Mother's Pride (bait for carp and me) and my trusty ten-foot Richard Walker Avon rod with Mitchell 300 reel, I stealthily made my way along the bank, peering over the high reeds. Very quickly I saw a couple of fish that made my heart leap. Two big carp, very big for the seventies, were cruising aimlessly along the margins into a corner. I could put my floating crust in and watch from a gap in the reeds, just around said corner, where I wouldn't spook the fish. I fired out a few loose offerings and gently cast my hook bait among them. Now, Mr Bond ... we wait. And wait I did from my little vantage point. But not for long. I knew the carp were in close, but a slight breeze had got up and pushed the pieces of loose crust further out into the lake, where one or two carp began to nose and take them! Surely any minute now my bait would disappear into a giant swirl of carp mouth. Except that my hook bait, courtesy of drag from my line, was no longer in the main body of floating baits, but still in the margins and a good way from where the carp were now feeding. Not only that but, just as I realised what was going on, I saw a swan heading right for my morsel of Mother's Pride at Mach 2 speed. No!

I ran to my rod, hoping I would beat the swan, when to my astonishment a huge mouth emerged from under my bait and sucked in the crust. Result: one disappointed and bewildered swan and one elated truant. I lifted into the carp and all hell broke loose. The fish was nearer thirty pounds than anything else and shot off into the lake towards a serious weed bed. I tried to slow it down but there was no hope, and we parted company. The howls of laughter and derision from the breadless swan were quite justified. Chastised for missing school and being seriously undergunned in the line department, I slunk home, vowing to be a good person in future and not play the hop again. I'm still trying.

Incidentally, the Mother's Pride mention leads me onto something I referred to earlier. It was bait for fish, but as a kid *a nice bit of crusty bread doubled as the best of all grub when you've eaten your packed lunch before you've even got to your swim in the morning. By midday even your hemp seed is look-ing tasty. Luncheon meat and bread is a banquet. No matter that you've been using the same fingers to hurl maggots out every cast. I don't remember ever having any gut problems after rummaging around in maggoty sawdust or mossy worms before tucking into food. Hand sanitiser was something only surgeons used.*

There is, of course, the school of thought that we need greater exposure to muck to build up our microbiome ... whatever that may be. Having said that, I did have a section of my colon removed – but not for another thirty-five or so years. I believe maggots are now used to treat certain infected wounds, and leeches have made a comeback, too. Oh, and by the way, I'm glad surgeons do scrub up. I wouldn't really want one to be mucking around with maggots, worms and halibut pellets before perform-ing a laparoscopic Whipple.

John: *Luck's a big deal in angling and Paul has made the point well. There's a saying that if your name is on a fish, then you can fish like a clown but it will end up in the net. If your name's not on it, you're doomed, no matter how good you are. I've seen it happen enough times to know there's a magical truth to this that I cannot begin to explain. If you do the right things often enough on the right water at the right time, success will come your way. Not necessarily straight off, but you will triumph in the end. This is a fact! I've seen grown men fail session after session and beat themselves unconscious with worry, but stick at it and success* will come eventually. The exhilaration will, of course, be much the greater.

But now the weird stuff! The more you obsess about a particular fish, the more you are likely to push it away. The more chilled you are, the more likely that fish is to come to you. This Bailey law was never more true than when I took groups to India for the mighty mahseer. I could even pick up the vibes at the airport check-in. Those anglers hopping around, fussy, nervous, talking incessantly, being a right pain in the arse, were almost always those who would fail. Conversely, the anglers who were quiet, attentive, helpful and avowed that the Indian experience was the thing they were after, nearly always conquered. You could say despite their laid-back approach, but I'd say BECAUSE of it. Often an angler only caught a mahseer of his or her dreams when they had given up, battered by heat, sometimes ill or simply exhausted by failure. Then, once they had opened up to the Asian karma, the mighty fish would begin to fall to their rods. Try your best, but keep a sense of proportion and luck will descend is my deep belief.

Another strange example. An angler can pike fish, say, session after session and catch next to nothing. The only luck going around has been bad luck, but then, out of the blue, everything changes. That angler is doing not a thing different, but fish after fish comes his or her way. It's like a miracle. All the action is on that lucky rod for a month, a year, even. But then, once again a switch flicked, luck deserts and the pike bonanza is a thing of the past. The cycle begins all over again and luck drifts away, hovering in the air above, picking where and on whom to next alight.

If we accept a level playing field in terms of skill, no one I have known has been lucky for a whole fishing life. Luck flits in and out, and the key is to fish your very hardest when you are on a lucky run. This is a window thrust open for you to exploit to its maximum, so work your damnedest every moment before it slams shut once more. You'll still catch fish afterwards, of course, but not in the same magical way as you did in your period of plenty.

Always bless your luck that you are there, alive, doing it. 'A bad day on the river is better than a good day in the office' is at last one cliché I have some sympathy with. There's always a shaft of light in any fishing day, a glimpse of glory that makes you realise why you go out to be wet, cold and blanking. It might be a sunset, a blur of kingfisher blue, the Milky Way painted on a night sky or even the smile from your fishing mate, but if you look, you will see and thank your lucky stars that you did.

CHAPTER ELEVEN

THE MAGIC AND MYSTERY OF FISHING

John: When I was ten or thereabouts, my parents bought me a copy of BB's The Fisherman's Bedside Book, *and I did just that, reading chapter after chapter in bed, hot milk and an apple close by. The book did me incalculable damage: a couple of chapters stole my heart – those by R. MacDonald Robertson and Charles St John, both Victorian fishers writing on the subject of ferox trout. The more I read about the power, majesty and rarity of this colossal predatorial trout of the great lochs, the more obsessed I became by the untamed waters they had inhabited since the Ice Age. I fell asleep listening in my head to the wavelets lap, the wild cats cry in the forests and those ancient reels howl to the plunges of great fish. But, of course, I had to grow up, progress up the angling ladder from the lowest rung I then occupied.*

I never forgot about the legendary ferox, so in 1978, just when I felt confident enough, I headed north to make my own history. For two decades I spent anything between fifty and a hundred days a year fishing in boats, in the shadow of snow-capped mountains, in storms, sleet or baking sun, and little by little I learned to catch these fish of dreams. Lochs Arkaig, Ness, Garry, Quoich, Fionn, Poulary and a dozen more … I feroxed them all, heedless of career, safety and everything but the chase for glory. I must have been half-mad (the greatest of angling writers, Chris Yates, told me I suffered from 'feroxide'). But, looking back, they were the best times of my fishing life: seeing those great fish emerge through the crystalline waters was proof that there was magic in the angling world.

Paul: *I know what John is talking about here – that added dimension, that great dollop of imaginative uncertainty that angling possesses.*

For me, the magic fish are probably salmon because they are big, beautiful and live barely credible lives in seawater and fresh. There haven't been many years of my life when I haven't thought now and again of Miss Ballantine and her sixty-odd-pound salmon caught from the Tay a century ago. Whenever I fish that surging river, I wonder if there could still be monsters like hers travelling under my boat, moving upstream to their destiny.

I've always loved the Red Lion Hotel bar in Bredwardine, Herefordshire, not just because it's on the banks of the Wye or because it serves great beer pulled by a great landlord; but also because there has always been a faded old photograph on the wall of past Wye monsters – salmon that defy all logic. Holy moly! They're two yards long and you can only wonder what they looked like forging up that river and whether we'll ever see their like again. Next thing I know I'm standing up to my bum in a freezing salmon river, and I've always got the whisper of these Goliath fish in my head. There's no other sport where the

unknowable plays such a leading role. Not that I know of,
anyway.

Winston Churchill was inescapably right about Russia when
he described it as 'a riddle, wrapped in a mystery, inside an
enigma'. But he could equally have been talking about fishing,
or at least *some* fishing. On one level, if you are fishing from a
woodchip platform, filling a net with sardine-sized silver fish,
the magic and mystery of angling aren't going to be a big deal.
But on another level, if you delve into the absorbing richness
of the sport, there's so much about fishing that defies rational
explanation, that isn't logical or sensible or provable, and
which plays to the heart and soul rather than the head.

In most sports, everything achieved is out there in the
public eye, but fishing can be more private, more subjective
and, perhaps, more open to embellishments. It's certainly true
that tales get swapped and grow in the telling. The mystery of
certain waters and the magical aura of huge fish inspire
legends that might be fantastical – or might just have their
foundations in truth. Anglers' bars, like Paul's Red Lion, have
been around for a long time, so drink might play a part; but
ego and the lure of fame have been part of fishing since
anglers first set up rods.

Apocryphal stories have dogged the sport for centuries.
John began his career on Sammy's Pit in Greater Manchester,
which was bottomless and, if he'd had the money for that
distance of line, he could have caught fish in Australia. His
nanny was born in 1888 and always swore that her fishing-
mad father kept maggots in his mouth during the coldest of

winter sessions. This, she said, kept them thawed out, wriggling seductively. (We've read this claim before but where's the cold evidence? John and his mates tried the trick in the dire winter of 1963 and were all sick for a week.) And has anyone *really and truly* tickled a trout? You know, winkled a wild brown trout out of a stream with their bare, unaided hands? We know it's all but impossible to get within ten yards of them with a fly, never mind trying to grapple them out with pork-sausage fingers. (Mind you, the prehistoric-looking Asian catfish called goonch are so dozy that local boys take an angler's line and hook, swim down and actually plant it in the fish's jaws.)

Pike, like ferox and salmon, have long been the stuff of yarns. Our twentieth-century angling heroes Richard Walker and Fred Buller both believed that pike of sizes way beyond human comprehension and experience lived in the vastness of Loch Lomond.

Walker examined the skull of the so-called Endrick pike, which resided dusty and forgotten in a priory on the loch's shores, and pronounced it to be the head of a seventy pounder. Around the same time, Buller hooked and lost a pike on the same loch, which Walker saw turn on the surface and, once again, guesstimated at a similarly vast weight.

Who can deny such apocryphal tales? It's true that way back, Loch Lomond was visited annually by runs of salmon and sea trout that might well have been fodder for pike beyond all reasonable expectation. But, whatever the richness of its environment, can pike truly attain seventy pounds? Well, they can in the Baltic Sea. There, leviathan female pike spend their summer lying well offshore around submerged islands, ambushing passing shoals of cod, herring and smelt. The fact that we simply don't know is where the drama of it all lies.

What about the pike's supposed ferocity? Old books tell of attacks on swimmers and dogs, and it's easy to regard these as fanciful. But there are other, more recent stories. We know a wildlife cameraman who kept a pond where he housed creatures to film: water voles, carp, newts, all the wonders of the water world. One day, our pal was out and John deposited a twelve-pound pike in the pond, a fish that was required for some BBC footage. John forgot that our man was working overseas at the time, and the pike soon melted from memory. Months later, our cameraman, home again, noticed ducks and water rats disappearing from the pond in mysterious circumstances. He thought little of it, until the day his four-year-old daughter started playing in the pond with a rake, which was grabbed so ferociously that in a plume of furious water she was pulled in and her ankle was seized by jaws of razor-sharp teeth. The battle was over in a trice, the shrieking daughter saved and the pike soon dispatched. But never tell us that a hungry pike is not a beast to be wary of.

Anglers themselves have been a mixed lot historically. 'Are all fishers liars or do all liars fish?' goes the old question. And while neither of us have ever knowingly told an untruth, some other anglers sadly have. A well-known angler back in the seventies notoriously went to Billingsgate Market in London, bought a fresh fifteen-pound salmon, painted it with black and red spots and claimed it as a rod-caught record Thames trout. One Norfolk Broads pike-fishing ruffian rose to infamy by claiming captures that weren't actually his and vastly exaggerating the weights of those he did land. To this day, his name remains a byword for treachery. Most appallingly, in the seventies a professional taxidermist nicked a thirty-pound pike sent to him, stuffed it and claimed it as his own record of forty-six pounds.

Indeed, there is no depth to which an evil angler will not stoop, and it is worth noting that Vladimir Putin has long been known to claim pike captures that are beyond credible. We anglers knew before anyone else the depths to which that man would sink. But bounding enthusiasm is, of course, different to downright lies. We love exuberance and anglers who hint at monsters around every corner. Leo Grosenipper was one such fisher whose dreams were always within casting range. His first trip to India saw him conquer the river on only his second day. We returned to camp to find bottles of champagne glistening in the lamp light and cigars for all. His guide, Rava, was feted and Leo clasped him all night like a brother. Next morning the chap was nowhere to be found. Leo had tipped him so handsomely that Rava had returned to his village to build a house. Yes, everywhere Leo fished he sniffed whoppers, and his belief drove us all to try that little bit harder, cast a little further, stay a little later. The magic of fishing is that you just never know.

Late last century, John was fixated on a lonely stretch of Norfolk's River Wensum where an eight-pound chub, a barely believable fish, was rumoured to patrol. There was also a tale of a gargantuan wild boar on the same isolated flood plain, a mythical creature perhaps deviously invented to put off would-be chub catchers. Undeterred by such an obvious ruse, John spent the winter there, alone in the elements, night after solitary night, catching decent chub but never the big one. And then he saw fable become flesh. The long grasses shook and parted, and in the light of his torch the boar stood, tall as a man at its shoulder, its eyes blood red, its tusks dripping blood. A yarn that John had discounted had proved itself horrifyingly true, and to this day he counts himself lucky to have escaped with his life (but not his tackle, which he left behind, to be collected next dawn). He never went back. The chub was safe and so was he.

We'd like to think this chapter sheds a light on the diverse delights that fishing offers you. On one level, a fishing session can simply be an indolent means to while away a sunny afternoon. At the other extreme, it can become an obsession, dominating your thoughts and dreams, leading your imagination along new paths, towards new quests. Fishing is a sport that can never be exhausted; it is always springing surprises and challenges, whether you are seven or seventy, such is the eternal fascination of what we do.

CHAPTER TWELVE

THE EXQUISITE BEAUTY OF THE CAST – OR NOT!

John: I can write this because Paul would be way too embarrassed to even mention such a thing. Now, don't misconstrue this, because I adore my gig with Gone Fishing, *but there are days when time hangs about a bit, when I have done what I can and it's all down to luck how things pan out. It's those particular shoots when I take great pleasure in watching Paul with a two-handed salmon rod doing his Spey casting bit. He's good. I think he was taught by Hugh Falkus, like I was, but in Paul's case the great man's genius rubbed off (not in mine!). In fact, Paul could be very good if he had more time to practise, and you appreciate his improvement as the filming goes on. His rhythm, his timing, his control – he just gets more and more in the zone. I sit and take it all in and realise once again how fishing, and fly casting in particular, is a real skill – a sporting art form. It's up there with a golf swing, a tennis shot, a Joe Root (or Ted Dexter) off-drive. Paul wants to catch a salmon, of course, but cameras or no cameras, I guess he'd have a great day on the water just unfurling that line.*

Perhaps the silkiest fly caster I've seen on the water was Jeremy Lucas, an England team member. Jeremy made it appear almost sexy – if it's possible for fishing to be sexy, that is. There's a feeling among many in fishing that prowess is all down to access to waters or luck, and that everyone in the sport is pretty much equal. This is annoying rubbish. Not every footballer is Lionel Messi. Not every runner is Usain Bolt. (Paul: I agree. I've seen Harry Kane many times, often at close quarters. He's immense, huge, powerful, Achilles-like. Imagine marking him for an hour and a half of purgatory. He's not like any Sunday footballer in any way, and, like John says, top anglers are on an equal level.) There are some anglers who, when it comes to casting, are simply genius. I count myself bloody lucky that I've seen a few in my time because I'm not even adequate with a fly rod of any type, single- or double-handed. I probably have no ingrained ability, but also, crucially, I was self-taught sixty years ago. I picked up all possible bad techniques and habits, and whenever I go out I'm trying to compensate for them. It doesn't matter whether it's fly casting, lure casting or thrashing out a carp rig a quarter of a kilometre, get instruction from the start and learn it right from day one.

Paul: *It's nice of John to say admiring things about my Spey casting, but I appreciate my limitations. In my defence, I only use the cast perhaps three days a year nowadays, and that's often when filming, when your mind is full of everything happening around. So, I'll never be effortless like the best fishers and I'll always have to think through what I'm doing rather than flicking out the line intuitively, but I can get it singing sometimes.*

96

I'll get that forward cast aimed up so that the line shoots out with such freedom you feel it clonk on the reel, saying, *Give me more, let me fly!*

I remember one morning on the Dee at Park, when Spey casting came together for me. I found myself on a pool successful for most but always a difficult one for me. As I stood there, a salmon head and tailed thirty-plus yards before me; it rolled on the surface in almost slow-motion – a glorious, stirring sight. Five minutes later I caught that fish and it gave me immense satisfaction.

First, I'd seen the fish and identified it as a possible taker. Then I'd known to wade out a little to get me away from the thick tree growth. At that distance, I'd calculated that I'd have to put out a long cast, and a near-perfect Spey did the job. I lifted into the fish at over thirty yards, which turned out to be silver and fresh run – the story-book end to an immensely pleasing scenario.

Like John, I was largely self-taught. I used to practise casting on a strip of grass outside my old tenement block in Hackney. I got a new fly rod on the day Spurs played Coventry in the 1987 Cup final, back when the FA Cup meant something. Spurs, unusually then but typically now, went on to lose. I would have been distraught, but I was comatose from booze by the final whistle. Since those days I have had the colossal benefit of being mentored. I've learned from exquisite single-handed casters like Marina Gibson and Charles Jardine, and from two-handers like the great Scott Mackenzie. I was also lucky enough to be taught by Michael Evans, another wizard with a salmon rod.

But like John, I have a hero and that was Hugh Falkus. We don't seem to have these legends in angling today. Big names *come and go, and it's good to have the chance to mention massive pioneers in the sport. Everyone knew Falkus could be difficult, but I think he was intrigued by me and took an interest in a young, slightly crazy comedian. One time I picked up a grilse from a pool on his own bit of Esk in exactly the Falkus Way, all stealth, unconventional roll cast and everything. Hugh was away on the Spey at the time, and when he heard he said, 'Paul Whitehouse, eh? A proper little bugger, that one!' Actually, he called me something much worse, but I guess he was delighted. He knew he'd taught me well and got me a catch I'd never have accomplished on my own.*

Safety is an issue when you're casting and most of the bad stuff we've seen has taken place when hooks are flying about your head. Hats are essential, but we'd never be without polarised glasses, because they let you see through the surface glare and, more vitally, they can save your eyesight from a fast-travelling hook. Check wind directions and speeds, and look behind you when you've moved fishing station. Is there anybody walking behind you? We've both seen bullocks hooked like this and it's not as funny as it sounds. Wading? Never go further or deeper than you feel happy with. Danger never goes on holiday. We've fallen in and hooked ourselves countless times, and calamity can come out of nowhere. Right, nannying over. What are the big casting considerations?

Practise casting on lawns or in parks. Paul did a bit of this back in the day and tradition says it's a good idea. Certainly, the more you cast, the better you become, there's no denying.

Accuracy is almost always more important than distance. Look carefully at where you want your bait, fly or lure to land before making the cast, and don't let your rod wobble side to side as the line goes out. Have a second of calm before making the cast and let the float, lead, lure or feeder hang still a couple of seconds before propelling it. And don't reel that float or whatever too close to the rod tip, but let it hang down a good couple of feet at least.

As the cast is made, what you want to impart to the rod is a smooth power and rhythm rather than exertion or aggression. You want to be in control of this important manoeuvre and **never** take your bottom hand off the rod butt because you'll lose accuracy, power, distance and you'll look silly. (*Paul: Bob often takes his* **upper** *hand off the rod during the cast and that looks daft and achieves nothing. And when beginning the cast with a fly, start with the rod tip close to the water. That is the LAW!*)

Check on the wind direction and current speed, and allow for these when you cast. Aim a foot above the water so that your bait or fly lands more softly. This sounds OTT but it works for us. Feather the cast before the bait's impact with the water. What this means is slow down its flight by gently touching the lip of the reel spool with the tip of your index finger. It needs a few minutes of practice, but it's worth it to see and hear your bait enter the water with barely a murmur.

Before the cast is made, check that the line is straight through the rod rings and not twisted. Have a momentary glance to ensure the line isn't caught around a reel handle. If there is a knot showing on the spool of a fixed-spool reel then it will catch the line as it flies out and seriously cut back distance.

Check that the weight of whatever you want to cast out there is heavy enough to achieve the distance you hope for. You're not going to cast a crow quill float a hundred yards, for an obvious example.

If you are fly fishing and the line isn't going out like it should, are you sure that the rod and line are properly matched weight-wise? It's no good having a light line with a heavy rod or vice versa. Make sure your back cast is working nicely, all straight and fairly level, before beginning the forward part of the cast. If you're letting your back cast get all droopy, you'll never get the power to make a success of the forward cast. Is your fly line travelling fast enough? It's never wrong to give it a bit of wellie and get that rod working and the line singing. You can overdo this, of course, but better to fail by being brave than being timid, we say! Are you trying to push too light a fly or fly line out into a strong wind? If you are getting into tangles, it could be that your leader is too long. Twenty-foot leaders have specialised jobs to do, but for starters a nylon leader about as long as your rod is a good guide. Don't be too ambitious at the start. Don't make too many false casts, as each one spells potential disaster. Better to get a fly out ten yards consistently and accurately than hit twenty yards once in a blue moon. Distance will come eventually; especially if your fly line is in good shape and not sticky or covered in sand or grit. Give it a good wash in warm, slightly soapy water and it will zoom out.

John: It is possible to be overawed by some of the professionals in the fly-casting game, mostly men who have a highly inflated opinion of themselves. I once wrote about Hugh Falkus that 'the sixteen-foot rod was a blur when he was casting, bending, cutting and dancing through the air. Forty yards of line were aerialised, arcing, floating and then arrowing over the water. He stood there like Moses with his staff, framed in spray and sunlight.' I even compared him to Sergeant Troy in Thomas Hardy's Far from the Madding Crowd, when he is flaunting his sword play around the novel's protagonist Bathsheba Everdene. Blimey, was I wowed or what? Intimidated, too – but I wouldn't be now. You just do your best, and if your casting isn't quite the best who really cares, if you are enjoying yourself and catching fish in the process?

CHAPTER THIRTEEN

A RIGHT MUCKING FUDDLE! TANGLES IN LIFE AND FISHING

John: I must have been the most tangle-prone child angler of all time when I wrote in my diary one day in August 1957: 'Went to canal fishing. Caught nothing. In tangles all day.' It's pretty stark, but the horror of it all is as raw as ever.

That August day, I walked down the cobbled lane and climbed the greasy steps to the Peak Forest Canal at around 2 p.m. I meant to fish till around tea time but got in tangle after tangle and went home early in tears. Mum was gardening as I stumbled up the path, sobbing my eyes out. She told me I was little and learning, and it would get easier as the years went by. I then had some crumpets and cheered up.

She was right, for the most part, and when I look back it was a miracle my mates and I ever got a bait in the water. The gear then was poor and mine was the worst. My rod cost 17/6d (87p), a price burned on my brain, and was nothing but two garden canes whipped with eyes and joined by a ferrule that stuck like it was glued in both warm and cold weather at the same time. The Bakelite reel was an abomination of a thing, designed to do *nothing but tangle, and that cost 1/6d (7p) — a waste of money even at that price. A further problem was that nylon line back then was costly and I could only afford to buy it in twenty-five-yard lengths. So, doing the maths, if a tangle lost me a yard or two of line, ten tangles could pretty near wipe me out. I had to try to save every inch if I wanted line enough to fish on till next pocket-money day.*

And the line too was rubbish. It was stiff, it had disastrous memory and could knot itself with miraculous ease. A cast of a couple of yards was the best I could manage, and my exhilaration was once cut short when my float flew to the other side of the canal and I realised the line had broken off at the reel. I was a mile from the bridge and headed back to retrieve it.

Were those days of trial and torment worthwhile? Of course they were. No matter how skilled you might ever get to be, a tangle is only around the next corner, and an apprenticeship of tears stands you in good stead to cope with whatever horror comes your way.

I remember in Mongolia in 2006, when two of my greatest friends Ping Pong, JG and I walked our long way to a camp on the River Shishged and spotted a colossal taimen hunting in the waters. Backwards and forwards the great fish swam, the whole metre and a half of it, harrying grayling, determined to make a kill. All we had to do was cast in. Ping Pong was first. His box-fresh thirty-quid lure sailed over the channel and buried itself eternally in a pine tree twenty yards into the forest. JG edged to the water's edge, cast and the brand-new line spooled out into an instant terminal tangle. I managed to throw out without releasing the bail arm and broke my line with the

*proverbial pistol shot, while the lost lure floated on the current
north towards the distant mountains of Siberia. Cocking up is
simply something that is always waiting for you, something
you'll never quite grow out of.*

Paul: *I'll admit that I've messed up endless times when I've been
salmon fishing, treading on lines, tangling with yards of
discarded mono, even breaking classic rods. It's just part of fishing: disaster always lurking round every meander.*

*As a kid, like John, it was tangling, not angling. I do remem-
ber with pain the day I lost my first carp, way back when all carp
were monsters because they were just so damn rare. The line went
crack (probably caught around a reel handle or something), and
the carp disappeared in a volcano of angry water.*

*I retold the story to my mates with something like pride. Way
back then, being broken by an unseen monster was a badge of
honour, a rite of passage. You'd brushed with a Moby Dick of a
fish, a creature of such untameable power that landing it would
have been an impossibility for anyone. That was the whole drama
of the case before carp had names and everything became so
professional and predictable. Mystery, monsters, broken lines and
rods – it's all part of a terrifying fishing journey. But let me add
that these days I would never boast of being broken by a fish.
Falkus called that a disgrace. We just can't escape the old bugger's
influence, can we?*

Good tackle, which is universally available these days, makes
tangles far less inevitable than in days of yore, but believe us,
it will always happen as long as people still fish. So, setting

out, what do you do to make life more comfortable and avoid the tears?

Before making a single cast, check everything out. Are those alder or willow branches above you going to cause a problem? Are the reeds and rushes growing tall in front of you going to be in the way? Brambles are the worst curse of all and nettles can be nasty. What about water weeds or lilies that could be sanctuary to a fleeing fish or could snag you on the retrieve? Are there obvious branches in the swim you need to avoid? What about the wind? Strong? Gusty? Blowing your tackle towards potential danger? None of this needs to take long. Just be aware and clock the natural world around you. Make it your friend, not your foe. And *never* sit or tackle up near a wasps' nest.

Before you cast a float, feeder, lure or whatever, don't reel it too close to the rod tip. Give yourself a good two or three feet of line and then, before the actual cast, just give yourself a second or two of stillness before moving into action. Always have a quick check that your line isn't looped around that top ring, too. Tangles are less likely if you are always in control.

Take your time. Don't do what John did in Mongolia. For example, ensure your line goes through *all* the rings and underneath the bail arm of the reel. Be calm, controlled, and if you start getting in a muddle from the off, and this happens to all of us, simply sit back, take ten minutes out and begin again from the start. Second time round, everything will seem more settled.

If you get in a bad tangle that looks more than a ten-second job, it nearly always pays to break off and set up again from the top. Line that becomes knotted can be weakened, and you can avoid that risk. Also, remember that if you are retackling, your line is out of the water and that can be a good thing. Fish are always aware of any gear in their environment, and by not fishing for a few minutes you relax them and make them more likely to feed. You might get a take the instant you cast back in.

Little things help. When you pack up, tuck your line in the clip on the spool of the reel. If it hasn't got one or you are using a centre pin, an elastic band around the spool can stop line flowing off and causing havoc in your tackle bag. Put floats back in a tube or they'll rattle around, get chipped and no longer be watertight. Be careful with pike traces. Count them out and count them in. Put all hooks back in boxes or rig tubes. They are better there than in your thumb. John is useless at putting rods in bags and tubes; he breaks three tip sections a year on average as a result. Broken tackle equals money equals work equals time not spent fishing.

Talking of traces, line and the rest, leave no angling or non-angling related litter behind on any account ever. We owe that to the environment and to wildlife and pets. Anglers have no right to get on our soap boxes and lecture others on how to behave around our rivers and lakes if we aren't squeaky clean ourselves.

On your perfect day, every cast will hit the bull's eye, you'll never get caught up, the wind will be still, you'll look graceful, serene, and not a swear word will escape your lips while you reel in fish after fish.

Forget it. It'll never happen.

CHAPTER FOURTEEN

FISH ON! WHEN TO WIND

John: Lost monsters hurt like hell at any age, but it's in childhood that the agony is intense and prolonged. That's how it should be, if the lessons of how to play big fish are to be learned properly. This lack of the pain of loss, if you like, is what Paul and I have against commercial fisheries where one decent fish follows another, and no capture or loss means all that much in a day of constant, easy success. Playing a big'un should be a matter of life or death, and the loss of a whopper should be a dagger to the heart, not just a shrug of the shoulders. Two lost fish, among thousands, scream to be told because the circumstances are as raw today as they were so very long ago.

The summer of 1964 was my eel era, when I was fascinated by these serpentine fish that still baffle science. North Norfolk was an eel Eldorado, and that year I caught them in the sea marshes, the dykes, the rivers and notably the estate lakes. I caught so many I sold them around the village, but it was the pelting madness of the takes that had me addicted – that moment when the cylinder of tin foil that was my bite indicator sprang from the grass and hammered to the rod butt ring while the line hissed from the reel.

Hooking the fish was good and the fight often pulsating, but it was that moment when a chunk of dead roach on the end of the line came alive that I lived for. And never more so than one broiling day around noon at Letheringsett Lake, five miles inland from the sea. There were a few of us fishing that day, our bikes lying in the long grass as we idled, listening to the harvest in full flow while swigging lemonade, having the occasional fight, all to the sound of one lad's radio murmuring the course of the Test match. My hero Ted Dexter made 170-something, and we were all cheering another four when my line rattled off and I found myself attached to an eel, the likes of which I'd never experienced before. The solid glass rod bent like a reed in the wind, and for every yard of line I sweated for, the eel took two. The leviathan, for this it had to be, was making for the island and the fallen trees, and soon it became obvious that, come hell or high water, I simply had to stop it.

The clutch was screwed tight as if by pliers, and the eel was forced up through the water, thrashing the surface to a foam. My mates were whooping, cartwheeling – and I was dead beat. The estate clock had chimed twelve as the line had scorched off, and now it rang out the half hour. There had never been a fish or a fight like it, and when the hook eventually straightened and the rod sprang back, I was devastated, like a weird umbilical cord had been cut and I was cast adrift in life. I've never quite recovered from that moment, although I did begin to sense for the first time that there can be honour in defeat.

Some years later, a pool close by was drained and an eel a mighty fifty-four inches long was taken from the muds. But mine was bigger still – a god among eels. I saw one of those mates of mine a year or two back. He remembered the event as clearly as I did. He said that when he heard the news of Dexter's recent passing, he first thought not of that most elegant of cricketers, but rather of that serpentine eel, lost all those years ago.

My second tale happened a couple of years before, and it taught me even more about loss. On a club outing to Oulton Park Lake for the day, there was the usual fishing match, on which we lads made little mark, and then we had a couple of hours to fish where we fancied. Some of us wandered towards the back of an island, and pretty soon my friend Geoff was into something way bigger than we had dreamed of before. It went this way and that, all the while steadily taking line with a heavy, unstoppable power. Soon we had a growing audience – all the adult anglers shouting (mostly contradictory) advice. Geoff, bless him, just hung on in there, all concentration, learning on the job. He was making a pretty good fist of it, and whatever it was began to look less invincible, until eventually we could even see the float and heavy boils beneath it.

The club secretary was the leader of the adult advisory panel, and he had got himself right by Geoff's elbow, telling him to do this, that or the other, puffing on his fag like a mill chimney. But here's the sad and disappointing end to the tale. As everyone's eyes were on the float, the man quietly removed the cigarette from his lips and, with obvious intent, brushed the lit end against the line, tight by the reel. Instantly, the game was up. The severed line whistled through the rod rings, the float and longed-for fish disappeared and Geoff dissolved into tears.

I pondered the man's actions to myself on the coach home, the next day and for a few years after. I never said anything about that mean, hateful action, not to Geoff, not to anyone. It took me a long time to process the fact that human nature can be good and generous, but it can also be bad and spiteful. It wasn't just an extraordinary fish lost that afternoon; for me, it was the loss of childhood innocence. Fishing. Life. They're much of a muchness to me.

Paul: *I go right back to my mid-twenties, when I was intoxicated with Welsh sea trout and a day on the Towy, right down on the tidal stretch. It was there, forty years ago, that I felt what it was like to be undergunned and taken apart by a fish.*

I remember I was using chub gear, a Richard Walker rod, probably an Avon – not a roach pole, exactly, but not serious enough for a fish that can run to fifteen pounds or more. I was freelining a worm, allowed in the rules of course, when it was hammered. It was a fish with the power of a barbel but with the inexhaustible energy of the sea in its bones, in every fibre of its being. This was a voracious predator by nature, but he was also back in freshwater with the spawning urge on him – talk about testosterone flying about.

Anyway, Whack! Off the fish set, and I knew what it was like to be on the losing end of a battle. What I did wrong was to try to stop the fish, like you would a chub or a carp. The thing is that those fish are fighting on home ground and know where to head for. A sea trout, or a salmon, isn't aware of snags and just heads for the salt. I could have let that fish run, sorted myself out a bit,

112

reset the clutch, drawn a deep breath and started all over. As it was, I held on and, Bang! Gone. Gutted.

Just as bad was a night on the Teifi. I'd read the sea trouters' bible by the one and only Hugh Falkus over and over as a young man, but I was undergunned yet again (never was one for learning). There was a moon up, not good for fishing, perhaps, but I was a bit of a novice and I welcomed the help it gave me. And I was right on the fringes of a village, with street lights shining on the water and a couple of cottages very close by. I was fishing a fly called Sweeney Todd, casting into this pool and retrieving with a slow figure of eight – classic stuff. I'd been at it a while when I saw this monster trout move into the pool and my heart stopped. I knew enough to expect the fish to keep moving, but not a bit of it and, blimey, what a take! The trout just went mental. It began to cartwheel like it was in a circus or something, and I remember saying out loud, 'Calm down.' I really did think it was going to wake up the entire village. Then, Bang! Gone.

I remembered Falkus's words about losing a fish being a disgrace, and I packed up and dragged myself home. I still feel wretched about that fish and the one from the Towy. Strangely, some years later I lost a thirty-plus salmon on the Dee. It was on for half an hour, right where my mates were assembling for lunch. It was a bit of a show for them, with a good amount of merriment, and when I manoeuvred the beast into a shallow, sandy bay, we all thought it was job done. Then the hook flicked loose and it was away.

The salmon, I never felt bad about. I'd have released it anyway and I'd had the drama. But those two mighty Welsh sea *trout: my head is in my hands remembering them those many moons ago. However, it could have been worse. I recall watching a guy playing a bonefish in a lagoon in the Seychelles. He seemed to be struggling, and he sat down in the water with the bonefish just a few yards away. There was a huge disturbance behind him, and a shark surged in and snaffled the bonefish from right between his legs. He lost the bonefish, but he could have lost a whole lot more.*

Losing a fish is a waste – of time, energy and effort. John has a lot to say about losing fish, especially after thirty years of guiding. Losing a wild, wonderful fish that you have set your heart upon landing cuts to the quick. It can be gut wrenching, demoralising to a degree that no non-angler can come near to understanding. It is not an overstatement to say that some famous lost fish have even blighted those unfortunate anglers' lives. You owe it to yourself to land every fish you hook, but also you have a duty to the fish itself. These are our guidelines to putting every fish hooked where it belongs – in your net.

Most gear is good these days, but cheap fixed-spool reels are still a very weak point. It's the clutches, you see; reels with plastic workings tend to have clutches that stick and are unreliable, especially in heat, cold and rain – pretty much all the time, come to that. You've got to check them constantly to ensure they haven't seized up, something they can do in a long fight. Our advice is to compromise if you need to on rods, but not on reels. John reckons that 90 per cent of fish his guided clients lose are down to clutches being set too tight and not giving line at critical moments. Set your clutch slacker than you might think to – it's easier to tighten than to loosen in an emergency. Or use a centre pin when the only clutch is your ever-reliable thumb! It goes without saying that your hooks are sharp, your knots are sound (even John's half-bloods!) and you are not fishing with line stupidly light for the job.

Always have a plan of action if you know that you are close to hooking a good fish. Think about what a big fish might do to evade capture and devise a strategy to counter this. Note any potential snags that could provide sanctuary and visualise how the fight might unfold so you are ready when it does. Your fighting technique needs to be right. Be bold as soon as you make contact, since you don't want a big fish to work up a speed that carries it unstoppably into danger. In fact, in our book, always play a fish with commitment and positivity. You'd rather lose a fish by being brave than being timid. Try to be boss from the start – though a big fish will have other ideas.

There are a couple of fighting tactics that are central to big-fish success. Side strain is up there, a concept that demands you move the rod either right or left, parallel to the water's surface, so you pull the fish off a potentially dangerous course towards lethal snags. Your gear is tough and powerful, so really lean into that fish of yours and you'll be amazed at what you can achieve.

Pumping is a big one, one of the essential arts of playing all decent fish. Start with your rod tip close to the surface and then steadily but gently ease the rod to the vertical, and the fish will follow through the water column. Keep reeling in and, once you are tight to the fish, lower the rod back to the original position and repeat the whole rhythmic process again and again. There'll be times you'll have to pause, and sometimes a real cracker will take line off you, but keep calm, keep at it and the fish will gradually tire – hopefully before you do. This is a tactic used by big-game anglers, but it is applicable to big barbel and carp, for example – think Paul on the Mole when he landed a clonking barbel for the cameras. Pumping supreme, that was.

All salmon anglers know that you can walk a big fish away from hazards, providing the fish is not alarmed (sometimes it's as though they don't even know they are hooked). Keep the rod high, put on a gentle, sustained pressure, and physically walk along the bank, leading the fish from rocks, submerged trees or whatever could spell disaster. The best example John can remember of this is his nigh-on lifelong mate JG leading a huge Mongolian taimen by the nose away from horrendous rapids that ran under a completely unscalable rock face. That mighty fish followed him upriver a hundred yards, never more than the rod length away, its big coal-black eye simply blazing unfathomable hatred. If that fish could have killed JG then it would have done so gladly.

Anyone who has watched *Gone Fishing* will know that Paul has a lot to say about winding or not winding, and the 'don't wind' exhortation has swept UK riverbanks everywhere. It's all about 'feel' – that instinct that tells you to give line when you must and retrieve line when you can. This comes in part with experience; the more decent fish that come your way, the better you'll be able to judge when to give and when to take.

But there is one big point to make here, the one Paul is always trying to get across to that maestro Mortimer: always give yourself plenty of line between your rod tip and a hooked fish. Probably getting on for a rod's length minimum is about right. The reason being that if a big 'un goes on a last-gasp run or plunge, then you have enough stretch in your line to absorb it without a breakage. Hold a fish too tight on a short line and, Bang, crash, wallop, you've lost it, Mr Mortimer.

CHAPTER FIFTEEN

AND AWAY – CARING FOR YOUR CATCH

John: An angling childhood is a torrent of tears – and not just for the big ones that got away. I remember a tiny loach that dropped off my line into a forest of nettles and perished, never to be found despite two hours of painful searching and much sobbing.

All of us have witnessed the death of immature perch, deep hooked, ravenously hungry for that tail of worm. Things quickly get better, however, and I don't think I have seen a fish die on me now for nigh-on fifty years, but that doesn't mean you can be complacent. You never take your eye off the possibility of causing a fish harm, and you come to realise that one of the joys of angling is seeing that exquisite fish power away unharmed, wiser and way harder to catch next time.

Now, Paul and I aren't fools, despite evidence to the contrary, and we do see why some might question why we catch fish in the first place. It's true that anglers are the genuine guardians of the stream and that fish in general would be much worse off without us, but that doesn't excuse any dereliction of duty and care. Personally, I don't mind being contentious, because every word of this book is the truth as we see it. I can see that commercial fisheries provide a safe place to fish and perhaps for beginners to learn – though I have my reservations on that one. What was it Roy Keane said about coaches in football? Something like 'I don't believe skill was or ever should be the result of coaches. It is a result of the love affair between the child and the ball.' It's exactly so in angling. I also question the mammoth catches, kept in nets and treated as commodities rather than as living creatures. Respect for these fish is at a minimum and that should never be the case. We might live in a supermarket-driven world, but the shopping-mall mindset surely has no place on a riverbank.

Second, and even less rationally, I have grown to query the catching of fish bigger than I am. This bit of nonsense started in Kazakhstan, on the Ural River, going back thirty years or so. Three Danish mates and I had permission to fish in a sturgeon reserve, and we were catching beluga of anything between 200 and 1,800 pounds. I say 'catching', but that legendary Danish angler and friend Johnny Jensen and I lost a fish that sort of size after a day-long fight. Those fish were abnormally huge, way older than us, and I began to think then – and have increasingly since – that they should not have been fished for. They should have been left alone. I repeat, there's no real logic to this, but sometimes your heart tells you what is right and what is not.

Paul: *Catch and release has been controversial at times, but done correctly it's hard to fault, especially with natural fish. Yes, you've got to do it right, and that means keeping the fish in the water, if at all possible. Certainly, when you are wading this is* *no problem at all, and we will map out all the rules in the next chapter.*

What I will say is that I relish that period when you have caught a big fish, often a salmon, in my case, and you are with it in the water, nursing back its strength. You might be cold, bordering on hypothermia, you might be lathered in snow, but it matters not a whit. There's this electric bolt that runs through you when you feel the rush of life return along those gleaming flanks and see the fins working again. It's more than a connection; it's like you drink in the essence of what a raw river is about. However, there have been times when fishing way out in the wild that I've taken a fish for food for the camp. Everything you look at around fishing is about survival in one way or another.

Pardon us if this chapter seems a little dry and like a check-list of 'dos' and 'don'ts', but the good of our fish is what counts. Tick the following rules off and we're proud of you. You can think of us as striving to practise what we preach. So, in no particular order …

1. Don't try to catch fish from places that are so snaggy you'll never land them. And think deeply about the most unconsidered of calamities. Twelve years ago, deep into the era when the River Wensum had lost most of its big roach, John located a small shoal of two-pounders. He got himself down there one brisk November dawn, and the first time he trotted his float down the river, it sank and his rod bent into a beauty. What a fish that was, deep-bodied, flashing silver and scarlet in the clear water.

He played it heart in mouth – it had to be two pounds, perhaps nearer three – and as he pussy-footed around, a pike the size of a shovel submarined from the reeds and clobbered it. The last vision John had was of a tattered and torn two-pounder being engulfed, a fate brought on entirely by its hooking. That fish had survived against all the odds for perhaps ten years, and within a minute of making its mistake, its life was snuffed out. John packed and left quite broken by the event. Lessons? Play a fish quickly and boldly for everyone's good. Be aware once again of that mantra 'danger never goes on holiday' and watch for menace around every bend of the river.

2. If you can, unhook a fish in the water without even touching it. This is where wading gives an edge. And *never* hold a fish up by its tail root, something we all used to see years ago. There's a real risk of dislocating the fish's spine.

3. Use barbless hooks or, even better, micro barbs. This is controversial, too, and we don't shy away from the subject. We've seen plenty of fish caught on barbless and their mouths can be slightly damaged when the hook slips, recatches and the process keeps repeating. A micro barb stays put and is dead easy to remove.

4. Always have forceps/disgorgers ready for use. If you are pike fishing, take the proper tools: pliers, wire cutters and scissors – and know how to use them.

5. If a fish is to be removed from the water, lay it on an unhooking mat or at least on wet, thick grass or a bed of water weeds in a very shallow margin.

6. If you must weigh a fish, use a large, wet, soft sling. Pre-set the scales before the fish is put in it. Enlist the help of friends if the fish is big, say twenty pounds or more. (*Paul: Personally, weighing fish just ain't for me. I just like to get 'em back.*)

7. If you want a photograph, try a portrait of it lying in shallow water. Have all the equipment ready to shoot. If a trophy shot is a must, support it gently and make sure the stomach isn't straining or distended. Put little pressure on the bone structure, and remember that the fish is cushioned by the water it swims in and when it is removed from its environment it is uniquely vulnerable.

8. When a fish is ready to be returned, support it in the current, in shallow water, where the pace is not too quick. Hold the fish very gently for as long as it takes to swim away confidently. Grayling are especially difficult to return. John has developed a technique where he lays the fish on marginal weed which is washed by the current. The fish is secure there untouched, and there is water passing over its gills. Ten minutes or more can elapse, but every grayling has eventually swum off without mishap.

9. For pike fishing, do you need to use trebles or will single hooks suffice? Have you been on a pike unhooking course? (Many clubs run them; if not, ask them to do so.) If bait fishing for pike, strike quickly as soon as a take manifests itself. (*Paul: I'd like to really emphasise this point. If you are unsure about what to do with a hooked pike, don't go pike fishing.*)

10. If you are using bolt rigs for carp, tench or bream, are you 100 per cent sure your rig is safe? We do not want a hooked fish towing a lead behind it and getting tethered in a snag somewhere.

11. We haven't even mentioned keep nets here. They might just have a place in the match scene; otherwise, we find them impossible to justify.

Looking at the broader picture, anglers should concern ourselves with the aquatic environment and protect it always. On a simple level, that means removing all litter and reporting suspicious incidents to the relevant authorities. More widely, think about joining working parties to regenerate spawning gravels and plant willows. Show support for the Angling Trust and bodies like the Wild Trout Trust. Paul and John are trustees for Fish in Need, a charity that issues grants to those attempting to make the life of fish better. Check FIN out online and help us if you can, perhaps? We are all brothers and sisters of the angle, as old Izaak would have said, and we are in this together, all trying to make fish and fisheries safer in the future.

John: To finish, a story. I was fishing on the Usk with a well-known fly angler. He was well downstream of me, but I kept hearing him say loudly, 'That counts.' At lunch I brought the subject up as we sat in the hut enjoying a sandwich and a dram. He told me he would snap the hook off at the bend so that when a trout took the fly it was momentarily pricked but could break loose after a split second, absolutely no damage done, apart from a little bemusement. My guy told me that his challenge lay in fooling a wild trout into taking the artificial fly, and after a lifetime of catching the things he didn't need to go through the rigmarole of playing them anymore. Taking catch and release to the ultimate, I'd say, but I liked the idea then and I do even more now.

CHAPTER SIXTEEN

OUR FISH

John: In mid-May 1990, my fishing partner at the time and I espied a massive gudgeon lurking under the bridge on the River Wensum at Elsing, in Norfolk. It lay for many days on the gravels there with fifty of so of its fellows, but not one came close to it in length or breadth. Back then the gudgeon record had stood for many years, decades even, at four ounces, four drams (which was how the record was always presented when I was growing up – and which is four and a quarter ounces). In new money, as people say, that's something like 114 grams, so not a monster in some eyes, but the most desirable fish on the planet in ours.

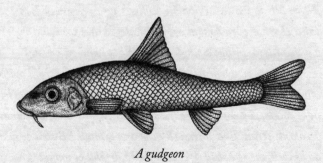

A gudgeon

I ought to say that the water was very shallow at this point, no more than two feet, and every scale on the fish's flank could be distinguished, along with the rays of its fins and the two barbules, or whiskers, hanging from its lips. A gudgeon is a mottled, brownish fish with an overlay of lavender blue to it that lends a certain sophistication, but in my eyes, and Roger's too, it was the size of her (I'll give her a sex now, out of politeness) that had us transfixed. I was teaching in those days, and after school I'd drive straight to the bridge, wriggle to the edge on my stomach and exhale with relief that she had still not been eaten or frightened off. She was still there, luxuriating on her gravel patch.

Of course, we could not make any attempt on her before the opening of the river season on 16 June, but six days before then, we began to introduce maggots into the stream from our view-point on the bridge. Initially our gudgeon girl held back, showing all the suspicion that had served (and saved) her well through her long life, but once she got the taste of the grubs, she was insatiable, driving away the smaller fish entirely. Only after she had gobbled twenty or more maggots did she became more relaxed, feeding only occasionally and giving up altogether once a couple of four-pound chub muscled their way upriver to join the feast. All in all, it was a campaign well set up, we believed, to the point that we even bought chemist scales so that we could weigh our 'record' to within a single dram. What could go wrong?

At 4 p.m. on 16 June, Roger and I met on the bridge with maggots as usual, but this time also with tackle – extraordinarily light tackle. A wand of a rod, a tiny reel spooled with 2 pounds breaking-strain line, a size 18 hook and a small BB shot would do us well, we reckoned. We could feed a dribble of maggots, wait till our girl became frantic for them, and then simply drop our hook, baited with two white maggots, over the side. The whole action would take place under our noses – we'd see every instant of the record-breaking action.

It didn't work out at all as expected. For three days she shied away, or small gudgeon got in first, or, on one occasion, a chub appeared from under the bridge and snaffled the bait a centimetre from her extended lips. But, at long last, the might of the two great anglers prevailed, and after a hair-raising tussle of a few seconds, she lay her massive body in the landing net. With trembling fingers, we hoisted her onto the delicate scales. We could not believe our eyes. We checked again – and again. Just under three and a half ounces (or three ounces, eight drams), not even 80 per cent of the record weight. Have I ever been more gutted? Well, perhaps. But not by much. I only tell this long-winded tale to make a point: ALL fish species have worth and charm. Fie on those trout snobs or carp obsessives, for they really do not know the full wonder of a fisher's life at all. Everything that swims is worth investigating, and that gudgeon, record or not, repaid every drop of effort handsomely. Never will I forget my three-ounce titan.

Paul: We have to go back to my mid-twenties for this one, when Victoria Park Lake in East London wasn't really known as a big fish venue. Nobody serious fished it, just kids for gudgeon and small roach, but I got intrigued when I was watching the ducks being fed. Every now and then there'd be a suspicious-looking black nose come up among them, and a piece of bread would quietly disappear. I kept going back to the lake to try to figure it out, to see what was there in all the scum and floating rubbish. A pod of carp, that's what, a gang of urban fish nobody knew anything about. There wasn't a lot known about carp fishing in those blissful days, and I tried to make a paste out of salt and vinegar crisps. I reckoned there were so many empty packets floating about, the carp would have got a right taste for them. In the end, I'd read about this new wonder bait, sweetcorn, and that's what landed me my nine-pounder. What a fish. My own personal carp, caught in this city jungle. I didn't go again. That nine-pounder was enough for me, and I was happy to leave the others alone and untroubled. I'd experienced my moment of bliss, a bit like a first love, perhaps. Sadly, those carp got discovered at last by other anglers. I guess I'd had the best of it when it was just me and them.

Yes, we love them all, and we like to think *Gone Fishing* has proved this, with targets from salmon to sea bass. Once, in Wales, Bob caught a minnow that might have shattered the record for the species had we weighed it. A fine fish indeed that caused a moment of mini-mayhem.

To a large degree, what you catch is a postcode affair. If you live in Rochdale, you're more likely to target roach than salmon or wrasse. And while this might seem bleeding obvious, it remains a major factor in what anglers fish for. It might do so even more now that we are entering an era of economic and environmental concerns: the car and the plane have transformed the locations we fish this past half-century, but the future might see us all having to be content much closer to home, which is all the more reason for us to back calls for vastly improved water management and care. We will need new fisheries and better fisheries, and they will have to be spread nationwide. Some species have suffered appallingly in our lifetimes, but there are glimmers of hope for a better future. A young John lived next to the River Goyt. In the fifties it ran the colour of whatever dyes the mills were using that particular day and not a stickleback could survive. Today, barbel and grayling, those prissy indicators of water quality, thrive there.

Talking of sticklebacks, it is not a bad idea to mention those days when a hook was a bent pin, when kids were content with fish two inches long and when the capture of a gudgeon represented a red letter day. Catching these 'jam jar' species kept both of us happy, along with millions of other lads and lasses. They served to give us an apprenticeship that, in our own particular case, lasts to this day. We won't go on about this, but we do want to make the point that starting out on monster carp or whatever is fine and dandy, but if you start at the top there's not much room to grow. We can get in a stew about ten-ounce dace from the Llangollen Canal even today, and we are grateful for that.

SALMON

Perhaps we'll start with salmon because it's just possible that they are Paul's go-to species these days. They are glorious fish, beautiful, exotic, bizarrely capricious and hard to read, as well as being lion-hearted in the fight. But Paul is a very good

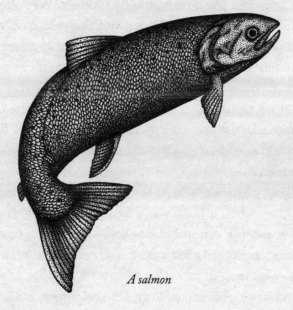

A salmon

example of how modern-day pressures have toyed with our lives. No one works harder than Paul, and his private fishing time, as opposed to filming time, is tragically limited. When, or if, he does get a free day, he might take it on a chalk stream like the Test because he knows the experience will be a joy. If he were ever to have a few days to himself, then he'd like to go for it, reach for the stars and pit himself against a salmon.

All anglers know that this century the number of UK salmon have dropped catastrophically, and we feel impelled to talk about the present-day fate of UK salmon for a moment here. Much of what we feel is immense sadness, a hollowness that the true king of fishes is facing such a bleak future. These are exceptional creatures with noble hearts, fish that lead buccaneering lives in saltwater and fresh. The fact that they

Casting a sunset lure as the autumn evening pulls in.
A fine time for predators.

Don't spook the fish. Let the heron be your inspiration.

The good angler is a nature detective. Hatching flies suggest
the water is warm and the fish will be feeding.

Sight fishing is as thrilling as it gets. John melts into the woodland.

Guzzling carp and tench turn the swim into a cauldron fizzing with bubbles. Fishing a float is the perfect form of attack!

Portrait of our beloved chub. A cracking five-pounder and a mouth that can swallow a sausage.

Paul fishing a salmon lie on the Tyne with superb control.

Freelined bread flake drifted down the current to a
group of big chub was exciting and productive.

John stalking grayling in the River Wye shallows.
The 'ladies of the stream' delight us both.

Paul battles a monster barbel on the Trent.

Cold. Fishless. Coffee calling!

'Can I have a go, PLEASE?' Paul and Bob on the lovely Welsh Dee.

Paul fishes an upstream nymph for a Tees grayling. You can almost feel the nip in the northern air as the winter approaches.

'What do you reckon, Bob? I can't even see the float!'

Simply exquisite. We anglers are privileged to
see fish as stunning as this perfect perch.

Nothing beats the experience of a float-caught barbel, especially
on the centre pin. Power. Beauty. Speed of a greyhound.

are born in river headwaters and make their way at a year or so old to the Atlantic Ocean, where they feed for three or more years, is fascinating in itself. But that they then retrace their journey over thousands of miles to spawn in their natal streams is beyond wondrous. Theirs is one of the great migrations in the natural world and we as anglers are in awe of them. Everyone, angler or not, should be equally inspired.

Our sadness that the salmon is facing extinction in many river systems is mixed with fury, a burning anger that we have allowed this to happen. It is true that there are cosmic factors, like the warming of the oceans, that have harmed salmon and are largely beyond our control, but that is a convenient excuse. There are endless things that could have been done but have not. We have allowed our rivers to become degraded by agricultural and industrial pollution. We have failed to build reservoirs and have instead abstracted rivers to their bare bones in order to service farming, industry and domestic usage. We have allowed raw sewage to be pumped untreated into our rivers for decades. We have encouraged trawler fleets to pillage the high seas for mature salmon with painfully inadequate quotas imposed upon them. We have allowed the damaging growth of salmon farms around much of our coastline, and the resulting excrement and diseases have devastated populations of wild fish. We have spectacularly failed to clamp down on the predators that harass juvenile salmon in the first months of their lives.

As a nation, we have turned a blind eye to the slow death of one of our talismanic species, and even now, more energy goes into reviving beaver colonies than saving our salmon.

There have been vibrant salmon stocks in the UK for millennia, and in our short lifetimes we have watched these salmon stocks dwindle and die. It is nothing short of appalling that this has happened.

Now, to return to a fishing perspective, we all know that salmon fishing is not what it was, but we've managed expectations and now know if you catch a salmon a day, then you have really achieved. Lack of numbers of salmon serves to increase the thrill of a single adrenalin-filled pulse of action, a classic case of less is more. The very best salmon fishing is probably found in Iceland (and Russia, before Putin's evil follies), and that has served to bring some UK salmon prices down a little and make that so-called sport of kings more accessible to the more lowly of us. Indeed, a favourite river of ours is the Tyne, and you can fish that very reasonably indeed – and even expect to catch a fish most days. Some salmon fishing on that inspirational river can be done with spinners. Flying Cs lead the pack, but in these days of enlightened fish care, perhaps take off the treble hooks they often come fitted with and replace them with a large single hook. We'll do anything to be kinder to our catch.

To *really* fish for salmon, though, you need to get some grasp on fly fishing, and that means learning to Spey cast to an acceptable degree. Far and away the best way of achieving this is to have a day or two of lessons, and teachers are not hard to find on the internet. Some of them can be a bit 'tweedy', but none are as frightening as that legend of the last century Hugh Falkus. Hugh taught both of us up on the waters of the Lake District and told John that he danced around like a marionette when he was casting and his only saving grace was that he could read the water in a reasonable way. Shorn of all confidence and permanently scarred, John fishes single-handed rods whenever he can.

Access to salmon water is easier than it ever used to be, again largely thanks to the internet. FishPal is a well-organised and accessible central booking organisation that is the salmon shop window for us all. Build up a relationship with the great folk there and you'll soon get all the gossip and up-to-date info and feel part of the scene. So, you don't have to be a lord of the land, a rock star or a business magnate to fish for salmon these days (though these occupations still help); good fishing is there for all of us with a quid or two to spare.

TROUT

Sticking a while with the high born, let's move on to brown trout, the prince of the river to so many of us. There are lakes with stocked browns, but the really classy stuff nowadays is fishing for *wild* brown trout, nearly always in rivers pure enough for them to breed. This is great news because arguably there are more natural brown trout around than there have been since before the Industrial Revolution – testimony to the fact that anglers have become environmentalists and worked hard to transform rivers that have been in a dreadful state for decades.

To catch a wild brown trout from a blissful, gin-clear river on a dry fly or a nymph is a wonderful thing to do. Again, we'd recommend lessons, and there are superb teachers around, like Marina Gibson, who appeared in *Gone Fishing* series four. The chalk streams of the south are the stuff of dreams, but there are magnificent rivers further north, east and west that are rougher, tougher and often infinitely cheaper. Olly Shepherd of Yorkshire introduced us to some cracking rivers around the Tees catchment in one *Gone Fishing* Christmas special. Most of us live within striking distance of a brown trout river (or lake), and that's especially the case now that the excellent Wild Trout Trust has rejuvenated so many stretches of water, urban ones as well. In the

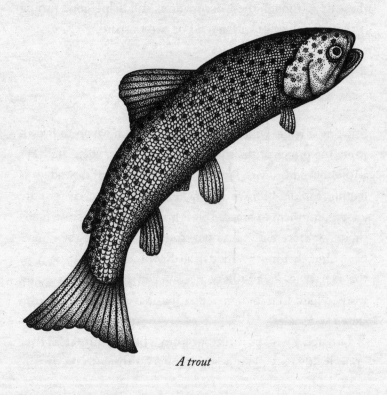

A trout

past week both of us have seen happy trout in streams as widely spread as Bristol, Leominster and Wandsworth. Much of this is down to the work done and the example set by the WTT, in our view one of the environmental bodies that actually *does* something on the ground rather than just talk in meetings.

Some brown trout leave freshwater and go to the oceans to live and feed, coming back to their natal rivers to spawn. These are called, imaginatively, sea trout and are as romantic and alluring as salmon to many purists – and to commoners like us as well. It is beyond a tragedy that many sea trout rivers anywhere close to the polluting presence of commercial salmon farms have been obliterated, largely by appalling infestations of parasites in the estuaries. Scotland's inspiring west coast has been hideously afflicted, as profit has been deemed more important than the environment, but many magical sea trout rivers have crashed in Ireland as well. Still, looking on what bright side we can find, opportunities still exist in the south-west of England, the Lake District, Wales and rivers in the south and east of England.

In his teens, John used to catch sea trout like yards of silver from his local Norfolk rivers, and runs of smaller fish can still be found here spasmodically in the summer months. A couple of years back, *Gone Fishing* went up to North Uist in the Outer Hebrides, and what an untamed, unforgettable experience that was. The message is look and you will find, and if you are lucky enough to do so, what a treat you have in store. A mild summer night. A darkening river. The splash of a hefty fish in the gloom. A heavy pull on your fly line and a screaming reel are excerpts from an angler's life that are impossible to better wherever your line is cast.

ROACH

Roach are something else altogether, a fish that both of us worshipped in childhood and still do today. Only a few years back, roach were the nation's favourite species in every poll taken, because they are pearls of nature, widely available and a joy to catch. Trotting a float down a river in the autumn or winter and seeing a stab of silver flank flash beneath the dark surface is addictive and accessible, and that's why there have been more roach anglers through history than any other type. Hampshire Avon roach. Roach from the Thames, the Trent, the Wensum, the Ribble, the Wye and even the Tweed have given more pleasure to more anglers than any other species, and probably still will into the future.

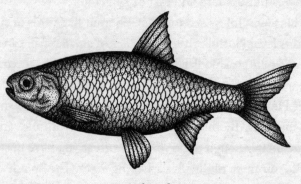

A roach

We'll give our roach tips later in the book, in Part Two: Our Special Sessions, but to be dead honest, roach aren't that difficult to make a start on with basic float gear and a pint of maggots. There are club and commercial waters, canals, pits, ponds and lakes in every parish of the land where roach swim. Once again, though, we have to mention that roach numbers and sizes are not what they were in the seventies, and while diffuse water pollution and abstraction have afflicted our rivers, the exponential rise in the number of fish-eating cormorants has battered many roach populations this century. Yet again, our statutory authorities refuse to face facts and deal with uncomfortable truths. But let's celebrate what we've got, why don't we? We can say, hand on heart, that whether you are seven or seventy, there's no better fish to catch and nothing can bestow more innocent joy than a net of roach.

BARBEL

Except, just possibly, barbel. In the eighties, John loved to salmon fish down on the Wye until stocks dwindled seemingly to the point of no return. But as the salmon failed, the river began to nurture an increasing population of barbel – a mere coincidence this, as barbel really have no detrimental effect on salmon, whatever the diehards would say. For John, the barbel have been some recompense, just as they have been for thousands of anglers on this river since the nineties.

You can catch barbel sitting on a box, watching a rod tip, but you can also float fish, bounce baits, wade, touch ledger

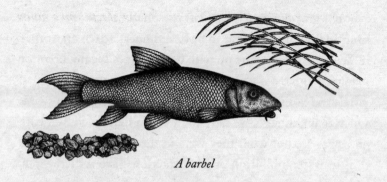

A barbel

and even fly fish for these amazing, beautiful, powerful fish that make your heart sing. There are barbel to be caught everywhere south of Scotland, in famous rivers like the Trent, Thames, Severn, Avon and Swale, and a plethora of stimulating streams besides. There are clubs, day tickets, syndicates; and there is the exceptional Barbel Society offering advice, waters and companionship. Do John and Paul have a favourite fish? Of course not, but what a blast a barbel can be.

TENCH

When John guided back in his Norfolk days, he realised that to very many anglers the tench is the finest of coarse fish, the queen of the summer, a prize beyond all others. Because he had access to some fine waters, tench-lovers flocked to him because they were starved of the species in so many areas of the UK. (*Paul: That's 100 per cent true. I have old pals who have fished every conceivable hot spot around the world for unbeliev-*

able fish, and all they hanker for is a float-caught tench on a sunny morning back in the UK.)

It's carp, you see: the present-day craze for big, bullish carp has in very many still waters driven tench to the verge of extinction, or at the very least to the point where they are

A tench

sadly dwindled in size, numbers and condition. In the sixties, finding a carp lake was quite something: sixty years on, finding one without carp is the harder task. A fine tench is a creature of wonder, especially float caught at dawn from a mist-cloaked lake when their feeding bubbles rise in sparkling trails and the lily pads rock to their activity beneath. When that float, red tipped by law since Izaak's day, disappears and the powerful, never-say-die fight rages, most would agree there is simply nothing better. Despite the detrimental effect carp have so often had, tench waters are still out there to be found. There are some clubs which understand what we are in danger of losing and work to protect this most endearing of species. Some tench are easy, suckers for corn, maggots, pellets or boilies. Others are pernickety to the point of impossible, and it's these tench teasers we'll look to conquer in Part Two, Our Special Sessions, later in the book.

CARP

So, to carp we must come, the species that has risen to dominate coarse fishing since the eighties. When we were growing up, riding a unicorn was easier than catching a carp. Carp anglers were few, men mostly, bearded, hermit-like tent-dwellers seen only after dusk and only rarely after dawn. We revered them as otherworldly beings. Whether they had actually caught carp or not was irrelevant; it was their holy quest that enraptured both of us. Just to talk to a carp angler, if we could find one in daylight hours, was like a pilgrimage of old,

A carp

a brush with wisdom way beyond our ken. Today, carp and carp anglers are found on every pond, their secrets open to all, the mystery exploded forever. This is a shame.

A carp is a big, dramatic bruiser of a fish that quickens any pulse, but there are carp anglers who demonstrate exquisite skills and fish by instinct rather than by stereotype. Go into any tackle emporium these days and you'll be assailed by every bivvy, bolt rig, boilie and carp-catching gadget that the wit of anglers (or manufacturers) can devise. You'll be advised to buy rods with the power and action of a broom stick, and to match them with big pit reels that hold a kilometre of line and have gearing to winch in a double-decker bus. But we'd advise caution. We'd suggest you go to a few waters and get a sense of the scene, see if it's for you. We might not be in thrall of them in quite the way we once were, but you'll read Part Two, Our Special Sessions, and see we still get our share.

PIKE

Pike are up next, the water wolves, the duckling eaters, the savages of the deeps. What we'd say first is get unhooking lessons from a club or anybody who knows pike dentistry (YouTube will do if it absolutely must), because it's no good pursuing a fish you are scared of catching. You also owe it to the fish. It's important that if any blood is shed, it's from your fingers and not from the pike. Otters and humans are the only threat in the UK to the welfare of pike, and while there is not much we can do about the former, we need to ensure our handling skills are perfected.

We'll also be quite honest in saying that small pike, called jacks in the trade, are a pain in the arse to catch. They wriggle, squirm and bite you if they can. Big pike, fifteen pounds or more, are a whole different beast. They are gobsmackingly majestic and lie docile if you treat them right. In Special Sessions we'll look at rigs and baits and approaches that catch

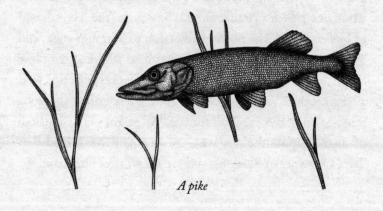

A pike

decent pike, and they work. Mortimer went out with us a while back and caught his first pike, weighing just shy of twenty-nine pounds. (*Paul: Should you ever meet up with Bob, he'll try to tell you that fish was a thirty-pounder. What do a few ounces matter here and there among friends, when the fish of a lifetime is the prize?*)

PERCH

How have we managed to leave it so long to mention perch, the brightly barred, sail-finned terror of every minnow in the land? In their blind passion for a tasty worm, small perch have been known to offer encouragement to many millions of junior anglers over the centuries, but once perch attain two, three or even four pounds, then what a sight they are. They have been called the 'biggest of fish' and you'll see why when you gawp at one of your own. Even these clonkers aren't too hard to catch – on worm, dead fish or lure – if you know

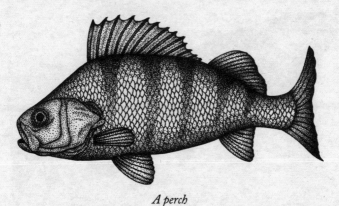

A perch

145

where to find them. In the seventies a scourge known as perch disease ravaged perch populations, but today our big rivers are a perch paradise. The Thames, the Wye, the Trent, and then you have vast reservoirs like Chew and Grafham, large pits and even commercial fisheries where striped monsters sate themselves on countless small silver fish. Perch anglers, we've never had it so good.

CHUB

Chub, too, have to be right up there in our favourite fish list because they are brassily bold, inhabit lovely rivers (and some not so lovely) and feed in any conditions – hot, warm, cold or freezing. You might fancy a barbel or big roach, and then a chub waltzes along and you think, Damn it. Don't. We've had a zillion sessions saved by a hungry chub over the decades, enough to ensure that we never disrespect them. And why

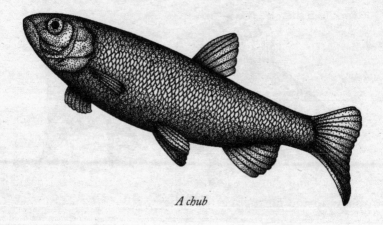

A chub

should we? These belters aren't just substitutes for another fish. They're not just some old timer you lever off the bench when the game's lost. No. They are mighty fish, champions in their own right, stunning and cunning.

HONOURABLE MENTIONS

Rudd: great plates of gleaming gold with fins a startling scarlet. **Bream**: horrid slabs of slime when small, impressive bronze shields when big. **Eels**: water snakes of slithering mystery – another fish species in a freefall decline that no scientist has an answer for. **Dace**: silver darts of fish weighing a few ounces – sixteen ounces if you are after a record. But at half that size, what twinkling sport of a frosty February day! **Crucian carp**: how we worship those golden, portly fellows, and how we'd love to see them back in every pit and pond and puddle like they were post-war. Crucian projects around the country show once again how angling and fish welfare work together. Be proud to be a caring angler, we say.

SEA FISH

Are sea fish for us? John used to catch **cod** off the Norfolk coast, great creamy white creatures that appeared in January and February, when the weather was rougher than a badger's whatnot and the winds came direct from the land of ice and snow to the north. The twenty-pound whoppers are gone

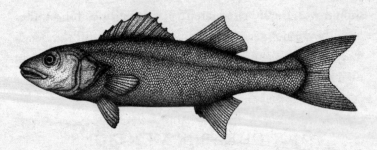

A sea bass

now, but would you know it, the proliferation of offshore wind farms gives hope of a return by creating vast artificial reefs that are trawler-safe.

Sea bass turn us both on, and why ever not? Sparkling, sea-bright fish, in the surf, off the cliffs, from the rocks, on fly, on lure, on bait, a species never to tire of. **Mullet**, so-called grey ghosts. Catch them if you can because we can't, though we love watching them as they follow the tides in water a hand's breadth deep and they keep our hearts racing with eternal but unfulfilled promises. Why, oh bloody why can't we catch them? **Pollock** are great, **wrasse** arguably better, caught

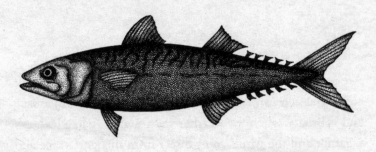

A mackerel

from rock citadels and pounding seas that scare the waders off you. **Mackerel,** the humble mackerel: not so much of the humble after one has broken a six-weight fly rod in half.

Fish. We love them all, really and truly we do. We urge you, if you are starting out, don't become a single-species obsessive. And if you are already in the game, open your eyes to all the extraordinary possibilities out there.

CHAPTER SEVENTEEN

FISHING MATES

Before we begin a chapter that is largely male-dominated, can we say that we'd love to see more women anglers fishing in the UK? For various reasons, far more women fish in the US and in Europe, especially Scandinavian countries, than they do here. But for many generations before there has been a noticeably condescending attitude towards women in the sport – and some would say there still is.

Years ago, there was a bit of nonsense about the 'female pheromone', the concept that the essence of femininity somehow rubbed off on a fly, lure or bait and drove fish, salmon especially, into some sort of frenzy. Nonsense. Women have caught huge salmon for decades because they fish hard and skilfully. John is extremely fortunate that his wife has become a formidable angler. Enoka was born in Sri Lanka and had no prior knowledge of the sport before marrying John, but she has proved to be as much a natural at it as any angler John has known. Both of us admire Claire Mercer Nairne, who is a tremendous force on the Tay, and, of course, Shirley Deterding might well be in her eighties, but she still fishes around the world. Her list of captures is phenomenal. Could it be said that had she been born a man, she would be acknowledged as the greatest angler of her generation? These are merely three names and, along with Marina Gibson and Lucy Bowden, the sport would benefit from seeing many, many more.

Paul: I don't like to highlight particular 'fishing mates', but Bailey has asked me so I will. I won't include him because it looks daft, but he's definitely on the list. Okay … he is my new best fishing mate! There are a few mates that I fish with regularly, but I can't include them all. Anyway, they should know who they are and not cry if they're not included. And there are others, brothers of the angle that I've not fished with for decades but who feel like they're constant companions.

My longest fishing mate and the one to whom I owe it all is my old man. My dad took me on my first fishing trip, and we fished together off and on for the rest of his life. He introduced me to the most glorious way to escape the world and be more part of it, simultaneously. How grateful am I? I can't begin to thank him enough. We fished for roach on the Lea, chub and dace on the Ouse, perch and rudd in various lakes in Ireland, but he was most happy on his native streams and rivers in Wales, like the Usk and Wye. Here we caught trout, grayling, chub and dace, though he was more interested in the trout. We fished Lough Conn in Ireland and Loch Awe in Scotland for trout, but we didn't fish for salmon till later in his life, when I started dragging him to Scotland. I think he felt it was the preserve of the rich and privileged and strangely 'beyond' him. But after going to the Dee a couple of times and fishing Commonty and then Park, where he was made to feel very welcome by Keith Cromer, the head ghillie, he learned to love those trips. Our shared love meant that we were together so much more than we would have been otherwise, and that has been precious. You can't put a price on it.

My oldest true fishing mate is Daryl. We have fished together since we were kids at the same infant and junior school. I was always envious of him and his brother (who I played in a band with in another life) because they lived right opposite the school. I imagined that they rolled out of bed, wolfed down a bowl of cornflakes and walked across the road to school in the space of five minutes, while I had to trudge through the estate for all of fifteen minutes.

In those far-off days of the late-sixties and seventies, almost every boy went fishing. We were no exception. We started to go fishing together when we were pretty young on the River Lea at Waltham Abbey, Cheshunt, sometimes Broxbourne, and also the North Met pit with varying degrees of success. But when we joined the Kings Arms & Cheshunt Angling Society (1973 junior champion, thank you) our angling knowledge began to develop … somewhat. Our club was affiliated with the London Anglers Association (LAA), and, having heard tales of great catches from other kids at Kings Weir, we began to get up at the crack of dawn on Saturdays to get the five o'clock train to Cheshunt. We'd schlep along the towpath for a mile and a half (it seemed like a marathon) at breakneck speed to get the 'chub swim' or the 'hemp swim'. Mind you, it was essential to stop and have a coffee from the flask and eat most of our lunch, before shouldering our boxes loaded with spam, hemp and Mitchell

300s (still surely the best fixed spool made) and half-running, half-staggering to the venue.

You had a key to get in and it was like opening a gate to paradise. Although quite heavily fished, it was a prolific venue and we had the boundless energy and devotion to fish hard and learn so much. Freelining for chub, hemp fishing for roach and getting the odd barbel … the experience we gained at Kings Weir stood us in good stead in later years with different species and different venues around the country and the world. Since my stent surgery I've been advised not to go to some of the more remote places where Daryl and our other mates go, so I've missed out on a few extraordinary trips, but we have fished together in Cuba, Russia, Iceland, Venezuela and the Seychelles for bonefish, tarpon and giant trevally. The most extraordinary places and the most extraordinary fish. Daryl and I will always fish together, till one of us departs for the other bank. I know he would agree that our groundwork on the Lea helped us to become better anglers. I'm not saying that I'm anything other than an enthusiastic and occasionally lucky amateur, of course.

There are people far better than I will ever be. People who work on the river have an eye and a knowledge that go beyond mere mortal anglers like myself. Watch Marina Gibson cast a fly, it's sublime. I'm not in her league. Or spend a day with my mate Jon Hall at Broadlands. I remember filming an Aviva commercial there once, and I was casting a fly. The crew were suitably impressed as my line floated out for the cameras. 'Wow!' they exclaimed.

'See, Jon,' I said to him as he watched with a smug demeanour. 'Brilliant casting.'

'In the kingdom of the blind the one-eyed man is king,' was his whiplash reply, without a pause. He is a truly brilliant angler. But Daryl and I have enough about us to be fairly confident wherever we fish, and it stems from those early days of friendship. And yes, yes, I'll organise a proper tench trip for us one day. It never fails to surprise me that the bunch of guys I've fished with in these remote and spectacular places really hanker after a trip for tench on the float.

Finally, of course, there's Bob. Daft Bob, cuddly Bob, infuriating Bob, 'Don't f***ing wind, you moron,' funny Bob, clever Bob and better-angler-than-it-seems Bob. Gone Fishing *became a big part of my and Bob's lives about seven glorious years ago. It came about very naturally while he and I were fishing together on a beautiful day on the Test. It's quite well-documented that we've both had heart disease, and some eight years ago I was alerted by my friend and old* Fast Show *collaborator Charlie Higson that Bob was having some problems. Bob's an old mate but he can be pretty 'reclusive', so I got in touch to give him the heads up and some words of encouragement about the stent procedure that he'd been told he needed. I remember Harry Enfield and I were just about to embark on our tour, and Vic Reeves and Bob were about to do theirs. I joked to Harry that Bob was trying to get an insurance payout and not do the tour. Anyway, I got in touch and reassured him, and in he went for a stent ... or two ... but his arteries were so blocked that at the last minute he had to have emergency open-heart surgery and a triple bypass. Very shocking. His surgeon later told him that if he'd gone ahead with the tour he would have died onstage (not for the first time) in Southampton.*

So, Bob then had a long period of convalescence, during which he was a bit reluctant to face the world. I badgered him and especially his wife Lisa, and I managed eventually to lure him out fishing. He had always wanted to go during the thirty-odd years I'd known him, but for some reason we'd never got round to it. Fifteen years earlier we'd got him a rod, then life and children intervened and again we put it on the back-burner. This time we might not have fifteen years! I took him to Broadlands, and he trotted a float and caught a trout and a grayling, I think, while I went off and caught a 21-pound pike on the fly with Jon Hall just to put Mortimer in his place. We then stayed at the lovely Greyhound in Stockbridge, and I think Bob was genuinely moved by the whole experience. A cliché, I know, but I believe it was a life-affirming trip.

We went a few times after that, and I remember one particular day he caught a trout on a mayfly and was made up. We were sitting on the bank mucking about and having a laugh; the Test looked like a jewel of a chalk stream, the trout were rising, the sun shone … it was as close to perfection as you can get, and it hit me. I said, 'This could make a good programme, you know.' My intention was to show that heart disease was conquerable with the correct treatment and that angling was one way (among many) of helping. Bob was also taken with the idea of exploring a friendship that had lost its way and how it could be restored. Anyway, we took it to BBC factual rather than comedy, because we're old and we had a factual element and we didn't want it to be out-and-out comedy. We had no script, just a few ideas: no voice-over, and we had Bob's 'cooking' and accommodation, about which I have no idea until we get to it.

Our first day was challenging, to say the least. Neither of us had done anything like this before, and we were stilted and wooden until I hooked a nice rainbow trout on the beautiful Derbyshire Wye; when Bob came with the net he slipped, and we both ended up on the ground in the mud but landed the fish. TV gold ... if only one of us had had a heart attack as well. That launched the show and we have gone on to do six series and three Christmas episodes with some memorable catches and beautiful moments. What a great fishing mate he is to have.

John: You can fish by yourself and for yourself, and many anglers famous and not so famous have adopted the solitary approach. It's not for me and Paul. On one level, I've seen the negatives the fish-alone philosophy inspires: greed, selfishness, paranoia even. Whitehouse and I are essentially gregarious creatures. It's not that we're overly chatty or needy, we just like human interaction of a meaningful kind. For me, a triumph is nothing unless I can share it with someone who understands and cares. Just the same with a disaster, too, and if you have genuine fishing mates they are with you every heartbeat. Everything you do is made more special by those attentive ears. It's give as well as take. We both love the act of helping friends to catch fish or to make plans about how to catch them.

Throughout my angling life, I have had big missions. Some of them have been bloody hard and, looking back, I'd have got nowhere without support.

It was Ron, thirty years older but still a mate, who taught me how to catch canal roach. It was with Pete, who had a car, that I started to explore rivers when I was twelve. Without Johns

Judge, Nunn and Wilson, I would never have had the success with East Anglian roach that I enjoyed. If Roger Miller hadn't partnered me in the eighties and nineties, I would have given up on Scottish ferox trout. Without Peter Smith, Wye barbel would probably have passed me by, and though I sort of wince to say it, if Paul Boote hadn't taken me to the Ganges, I would have missed out on twenty-five years of Indian mahseer fishing. Would I have spent so much time in Mongolia without Czech friends? And today, even, would I fish at all without the daily interest of Pingers, Ratters, Clarky, JG and the rest? And what about Whitey? Every day throughout this winter we've been writing this book, and I've been at my desk or down on the river, where his inevitable call is something I wait for and even need. He might be two hundred miles away playing Grandad in Fools and Horses *but he's everywhere with me in spirit, and I rely on that.*

I don't mind if this sounds a bit weepy – it's how it is. We've always said we're only writing what we know about, and we both know fishing is made a million times better with mates.

CHAPTER EIGHTEEN

OUR FISHING FUNDAMENTALS

We've relayed quite a bit of info so far, so now that we're at the end of Part One, we thought we'd distil what we consider to be the essential truths of fishing – our Fishing Fundamentals, if you like. You can flick to this page whenever you need a reminder of our philosophy – or even some inspiration to get out on the water.

Enjoy – and see you on the other side of the bank!

1. **Look after our fish.** We all have to do what we can to look after our fish and our fisheries; they are our everything as anglers and as human beings. Catching a fish should be a big deal for all of us, and we owe it to every one of them to appreciate them and treat them well. Let's not create carnage in the world of fish – let's leave them unmolested as much as we can.

2. **Look after our waterways.** It goes without saying (but we're saying it anyway) that we don't leave litter – and we pick up other people's while we are at it – and that we

follow the Countryside Code. (*Paul: Memo to self – look up the latest version.*)

3. **Understand your fish.** In our fishing universe, we always put the fish first, way ahead of bait, tackle, rigs and all that stuff. Set yourself to understand fish. Watch fish. Think why a fish does what it does. Realise that a fish thinks almost exclusively (we guess!) about survival, food and reproduction. Nothing else comes into their equation, so time is *never* wasted in getting close to your fish. In fact, it's the reverse. Matt Busby famously said it only takes a second to score a goal, and the same is true of catching a fish if you have laid the foundations right. (*Paul: Actually, John, it was Cloughie who said this, not Sir Matt. You're a typical United fan who thinks your lot has invented everything good in the game.*) You've probably picked up on our passion for Izaak Walton, which comes about largely because all those years ago he immersed himself in fish behaviour and studied it as a way to catch them. Of course, in the mid-seventeenth century he couldn't rely on fancy gear to do the job for him; he was starting almost from scratch and that meant fish focus. It's exactly why we revere Bernard Venables, whose cartoon hero Mr Crabtree took his son out with him and taught this nature-first approach that we learned as kids and have lived our fishing lives in thrall to.

4. **Make that first cast count.** Please accept that you don't rush to get a bait in. First you find the fish and then strategise an approach. Make your first cast only when you have weighed up all your options. Make it really

count. This might be your best, your only chance, so get everything as right as you possibly can. Don't waste that cast sploshing out a random feeder, lead or float, but put a bait in as though you really mean it.

5. **Location, location, location.** Right place, right time, right bait has been the rule forever, certainly since we were kids. The first bit is probably the most important, so think what you want to catch and hunt out where you can catch it. Tackle shops (sometimes), chats on the bank, internet searches for local clubs and day-ticket waters are all potential sources. Sometimes it's good to jot down every scrap of intel. A bit nerdy, perhaps, but John has kept a notebook most of his fishing life and many is the time he has been glad he's done so. Ordnance Survey maps have guided us to some of the best places we've ever dreamed of fishing, and guess what? Even in the days of Google Maps, those paper treasure troves still serve us well.

6. **Get the right gear.** Always make sure your gear is strong enough for what you want to catch, that your hooks are sharp and your knots are good. Map out where snags are and visualise what you will do in an ensuing battle. Be brave. Trust your gear and your skills.

7. **Less is more.** Let's say you have caught a good fish. In every way it's a good idea to rest the water and let the fish (and you) recover. Go for a walk. Have a natter with a mate. The longer your line is out of the swim, the more the fish will relax. And don't always assume that two rods give you twice as much chance as one does. Often, less is more, and two (or three) rods halve your chances, not double them.

8. **Study the water.** As you walk, watch the water closely. Swallows and swifts wheeling over the water might mean a fly hatch, which could mean fish-feeding activity too. The sight of rolling tench, for example, confirms this. Is there any stained water you can see? If there's a patch more coloured than what's around it, that means there are fish there, feeding off the bottom and throwing up the silt.

9. **Listen to your instincts.** It's good to have a day worked out in your head before you even arrive at the water, but if you see feeding activity then it might be better to scrap your first plan and move onto another. Be flexible and keep an open mind. Sometimes it's good to move ... but sometimes it's better to stick where you are. This is one of the great enigmas in fishing, and you have to work out the pros and cons carefully ... and listen to what your instincts are telling you, too.

10. **Enjoy it or go home.** Stop if you are not enjoying it. It's a joy, this fishing lark, and you're not trying to put food on the table for your starving kids. A famous angler, Fred J. Taylor, was fishing in the ice and snow and said, 'I'll be glad when I've had enough of this.' There's no shame in going home. If you are bored, then you are not working at it. When John is guiding, or when *Gone Fishing* is being filmed, he and Paul have a succession of plans. If A doesn't work then he moves to B, and right down to Z if he has to.

11. **It's only a fish!** Never begrudge anyone else's success, because it's simply not worth it. Learn from their success

and take from it what you need to improve your own performance. Never despair. Do the right things and your time will come. Don't get down in the mouth – it's only a fish. And always remember what a privilege it is to be alive and somewhere lovely – whoa, there's a kingfisher going past! See what we mean?

PART TWO

HOW WE FISH IN PRACTICE
– OUR SPECIAL SESSIONS

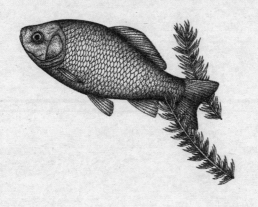

So far we've talked a lot *around* fishing without going too deep into the mechanics of it. Now we'd like to reflect on our fishing lives, describing sessions that have been special in some way and analyse what we did right or wrong. We'll suggest exactly how you might go about similar situations – and quite probably do better than we did. If you are fans, a few of these chapters might be familiar to you from *Gone Fishing* episodes, and they might give you a little insight into what went wrong or right in front of the cameras. Most of the sessions, though, are about us on our own, offering insights into how we have approached challenges throughout our angling lives. What we hope is that these tales inspire you to create your own angling adventures. To help you on the way, Part Three: The Nuts and Bolts of What We Do will fill in some of the more precise details of how we both fish and approach our challenges.

We're not in any way judgemental (much!), but you won't find much in the way of self-hooking rigs or non-natural waters. Let's explain. We *are* old school, but there are many of us who think the same. Our friend Chris Yates, the one-time carp record holder, labelled the bolt rig the death of 'proper' fishing way back in the eighties. In so many ways he has been proven correct. It is possible to tour every water in the country with a boiled bait (poetically abbreviated to boilie over the years) and fixed lead and catch every fish that swims. Many anglers of this persuasion might even be asleep at the time. This is not to say that all long-stay anglers are wedded to one-dimensional approaches, of course, and many carp anglers are expertly innovative, we know.

But we are wedded to action. Our sessions are hands on, tactile, often mobile and constantly evolving as we watch what the fish and conditions are doing. We probably fish shorter sessions than some, certainly not week-long stays, but they are action-packed and we are always strategising to make our chances better. 'Strategising' is too posh a word for what we do, but we like the sound of it, even though 'thinking around stuff' is nearer the mark.

We fish like this, just as so many others do, largely because of time. Like everyone, we find there just aren't enough hours in the day. Another factor is crucial: *how* we fish (that phrase again) is very much more important than *what* we catch. Paul would rather catch a single brown trout on an artificial fly than a dozen on maggots. John would rather catch one barbel on a float than a dozen on a feeder. We don't regard this as elitism, and in John's case he'll use a feeder in a six-foot flood if he has to. It's just that we are not fishing to put food on our family's table but for pleasure, and this is how we find that pleasure. If it works for us, we hope it will work for you.

We will start with Winter Sessions and then work through the seasons in order, introducing some of our favourite fish to look out for during each season. We will look at great sessions that have enriched our fishing lives, perhaps when we have made *Gone Fishing* memories or just spent time in each other's company. But remember, we are about simplicity, and in these sessions we'll be concentrating on the fish, the water and how we can use our watercraft to achieve a result. We're not saying that the rigs and baits have no importance, just that they are not up there in first place. And, remember, for us it is all about enjoyment and keeping the wild fires of enthusiasm burning. We don't do dull, we like to think; fishing is just too captivating for that.

CHAPTER NINETEEN

WINTER SESSIONS

Winter, we love it – though not the ceaseless, icy rain that carries traces of snow. Northerly winds are desperate, but easterlies are worse and Beast from the East conditions are as bad as they can get (for us and the fish, too). John always says that the *only* time he finds wild fish killed by otters is during these prolonged periods of extreme cold when fish go into semi-hibernation and can be picked off by our furry friends. Chub seem especially defenceless when they are comatose like this; but, of course, when the water is warmer and the chub are up and running, the dynamic changes completely and otters struggle to survive.

In winter we are looking for periods of stability and calm. We like high pressure, if the nights aren't too cold, and those exquisite Christmas card mornings when the trees are frost-laden against Mediterranean blue skies (perfect for grayling, too). Dawn can be excellent – and dusk especially at this time of year, more so when there is little or no wind and the water reflects the sunset like a vast, glowing mirror. But what we really long for are those periods of mild westerlies when there is constant cloud cover, intermittent drizzly rain, steady Force 2 to 3 winds, no frost and daytime temperatures of ten or even twelve degrees. Celsius, of course! Funny how all us old boys talk pounds and ounces and miles and yards but have gone all metric over temperatures. The vital element here, however, is that when you get these weather patterns, you know you have to capitalise if you can; it is fishing gold. During winter fish are mostly active early and late, but when the westerlies set in, any time can be feeding time.

Wind has a big impact on winter fish. Forget easterlies; but westerlies and southerlies are a different matter. If these are prolonged and brisk, they'll set up underwater currents in still waters that really stir fish into life: bream, roach and even tench if it's blissfully mild. (*Paul: A winter tench really sticks in my memory. I remember it like an old friend turning up unexpectedly – one you really want to see.*) A good blow also colours water a little and that stirs up food, which is a bonus, but it is the security issue that is paramount. Once that crystal-clear water is clouded, even if only by a little, silvers in particular feel a tad safer and are happier to stray from cover. Never forget what a tough life it is down there for a half-pound roach that everything else wants to eat. It's nice to be wanted, but not quite that badly.

And what about floods? John has the Wye as his home river and knows all about flooding. It only has to drizzle in the Welsh mountains for the levels to rise six feet overnight. If it pours, the Wye valley becomes a vast lowland lake. If rivers are within their banks, though, it always pays to fish.

The key thing is not to be afraid of high water and to realise that it can be on your side. Chances are, rain means westerly winds, which mean warmer temperatures, a factor strongly on our side. A rise in the river can push fish into those big slacks you find along the margin of any river. Here the fish will shoal together and will often be hungry if they have not fed for days. You might need to use slightly heavier gear and also decide on bigger, smellier baits. (*John: I use pieces of luncheon meat the size of my fist when the Wye is roaring through.*)

Both of us adore the Wye with a passion, but we accept that she's a tricky piece of work after November. She can be up and down faster than a fiddler's elbow, and big, rain-belching clouds over the mountains of wild Wales can spell disaster for our fishing plans, particularly for Paul, since he has a trek and a half across the whole breadth of the country to get there. (*Paul: That's true, but even though I hate that trek east to west across England, at least in a flood you'll often know where to find the fish. We've already mentioned that the slacks can act as havens in high water, and they are the obvious place to start … providing other anglers aren't about, thinking the same bloom-ing thing!*)

We have a well-worked-out plan to tame the Wye's volatil-ity a little, a plan that might save you time on your rivers as well. And it can all be done before you set off anywhere with your rods. The first thing to do is jump online and check the Wye river gauges, updated constantly by gov.uk's 'Check for flooding' service and the Met Office. For the Wye we look for the levels to be between two and three metres on the scale and falling. If she's at three metres and rising, Paul stays put in Islington. Then we watch what's happening with the rain-fall upriver, at nearby Builth Wells and Rhayader, where heavy downpours lasting more than three or four hours will mean flooding carnage on the middle Wye, where we want to fish.

We also have a ten-degree rule: we add together the mini-mum temperature by day with the minimum temperature by night. If the total exceeds ten degrees we're in business; but if we see five degrees by day and a single degree by night, for example, Paul's car stays in London.

Floods are all about low-pressure systems, but what about high pressure, frosty nights and bright daytime skies? Often it pays to fish lighter in such conditions than you would when fish are feeding much more enthusiastically. When everything slows down and the temperatures begin to tumble, a tiny size 20 hook baited with one maggot will often catch fish that are reluctant to take three maggots on a larger size 14 hook. Delicacy is all when waters are cold and crystal clear. (*John: Back in the seventies, when I was big roach obsessed, freezingly cold conditions sometimes forced me to use tiny hooks and lines seemingly no thicker than a spider's web.*) This doesn't always have to be your approach, but it might well be in bright sunlight, early morning after a hard frost. When fishing is *really* tough and if you are well rugged-up, try fishing later than you might otherwise like to. That hour into true darkness can produce bites that seemed impossible earlier in the day.

What do fish do in the winter? They are probably not feed-ing as hard as they might in summer, so cut back on bait (though carp can be hoovers if they are in the mood). The water is generally more clear in the winter, as algae blooms die off, so fish are all feeling more exposed to predators; especially as winter is cormorant time, which visit from Eastern Europe in their thousands. Fish of all sizes seek cover because of predation. Look for them close to died-back weed beds and deep in reed beds.

There's a lot to find, come winter. Perch can be on fire and look great in clear water, which enhances those great black bars of theirs. Bream will feed day-long when wet westerlies set in and the temperatures are ten degrees plus with mild nights. Head out into choppy water where the breeze hits and use plenty of smelly bait: hemp, corn, casters, mini-pellets, a few additives in the groundbait, and you're in for a good day.

When we were kids, nobody thought of winter carp fishing, but we all do now. They're unpredictable creatures and we've seen them behave like you wouldn't expect. John has watched them feed day-long under ice an inch thick and in water only two feet deep. You'd expect carp to come on the feed late but not always, and a bright sun can get them fizzing at midday when the light is at its strongest.

OUR LOVELY LADIES OF THE STREAM – GRAYLING

Grayling are stunners and sum up what we both love about the winter. We like them so much because they live in great-looking rivers and because they aren't quite game fish and aren't quite coarse fish. You fish for them from 16 June to 14 March, which makes them coarse in their spawning habits, but they share water with the lordly trout and salmon, as well as an adipose fin. They're a bit like we feel sometimes in exalted company, the poor relation somehow … but we get on with it, and nothing comes between us and our love for grayling fishing. (*Paul: Mind you, we are not alone. I'd say that in recent years grayling have become as valued as trout on many rivers – if not more so.*)

Another bonus is that you can bait fish and fly fish for grayling with equal aplomb. Nymph fishing is a delight, but trotting a float seventy yards on a chuckling Yorkshire stream is seventh heaven. It's not right to dismiss grayling, as some do, for being hard to find or inordinately expensive to fish for. Grayling are common in the rivers of southern Scotland, in northern England, in Derbyshire, throughout Wales, in parts of Devon and all along southern England, from Dorset through to Hampshire. Many trout fisheries are wise enough in these financially challenging times to offer grayling tickets from the autumn, when trout fishing ends. Really, there's no excuse for us or anyone else to ignore these wonderful ladies of the stream, as the old angling books dubbed them.

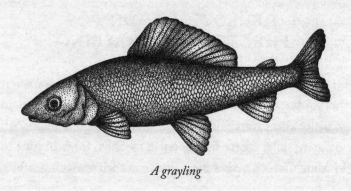

A grayling

Grayling have been a favourite with *Gone Fishing* for a while; we've had three outings for them, including the Christmas special on the Tees a few years back. Despite our location on great stretches of that river, we had a shocker, just as we did on the Ure a couple of years earlier.

On the Ure, John fished a day on the water before the main crew arrived. He was seriously discombobulated by the passing of his old mate, TV star of the eighties and nineties John Wilson that very morning, but it didn't stop him concentrating on the job in hand. His investigations revealed a particularly lovely, smooth glide of water, which absolutely shrieked grayling triumph. Grayling love a steady push of water, not too deep, over clean stones and gravel. All around the river there was beauty, peace and a bank of oaks turning to gold. It had all the signs for success and it could not have been more accommodating for filming. The shallow margins allowed easy camera access so that the crew could get close-ups and all the angles the director required.

John had enjoyed an hour at the head of the beat, trotting maggots, and had picked up a grayling of a pound and a half and a couple of around a pound in short order, so confidence was high on every count. Even the weather was on our side, recent rainfall having left the water with a perfect tinge to it. John had left the swim full of fish, so what could possibly go wrong?

Bob fished the lower section of the water with a 13-foot trotting rod, 4-pound line, a stick float carrying 3BB shot and a size 16 hook baited with three maggots. The trick is to feed steadily and generously every cast, to let the float trot the current and make sure it keeps true to its course by mending the line. He waded out nicely into the flow. There was nothing wrong with his fishing position: there was no wind, so his control couldn't be faulted.

Paul stationed himself upstream, fishing Czech nymph style, and worked the whole sixty-yard run down to Bob meticulously. John watched both men at work and willed them to succeed. Neither was doing anything remotely wrong; in fact they were doing a great deal of what was right. Trout came along periodically, but the grayling absolutely did not. For three whole filming sessions the boys worked their socks off until, right at the death, Paul hooked a snorter of two pounds and more, which rolled away at the net. (*Paul: Mind you, if Bob had picked up the long-handled net rather than the pan-handled one, it might have been a different story.*) John was waiting for a hernia operation at the time, but the pain in his groin was nothing compared with the agony of seeing that fish right itself, flex that glorious fin and drift imperiously away in the current.

We all drove away puzzled, exasperated, looking for answers. Could it have been the crunch of camera operators' boots on the loose gravel? Or perhaps an otter had taken up residence close by? In the post-match conference, both of us surmised that perhaps John's test session had pressured the fish into moving; grayling are wild fish, after all, alive to the slightest sign that spells danger. But they also live in generously large shoals, and we would have bet on there being a couple of score of fish left. Paul made the good point that during our filming the river rose by a foot or more, and neither of us have had much luck when the water is mucking up like this.

The big lesson to take away is this: choose a good swim and fish it with complete confidence. Feed conscientiously and don't stop – it can take an hour for grayling to respond to a steady flow of maggots raining past. If sport is slow or non-existent, keep doing the right thing and the swim will switch on. Or it should, if you have faith.

But here's the thing that continues to haunt John. Below the run Paul and Bob were fishing was another hundred-yard straight that was deeper, faster, harder to fish. Was it possible that the grayling had dropped down to escape the commotion above? This is one of angling's greatest conundrums: how long do you give a smashing piece of water before you give up on it? How long do you stick with it, and when do you decide to move? To this day, John blames himself for not suggesting the downstream swim, certainly for the last filming session. There are lessons for us all to heed, and while it is great to have a plan, go off-piste if the initial one is clearly not working.

The Tees, where *Gone Fishing* filmed in late October 2020, didn't prove much more successful for us, despite the colossal input of Olly Shepherd from Fly Fishing Yorkshire. (How wonderful it is to find a guide and friend who will sweat blood for you. Find such a person and place your best fishing days in safe hands.)

Once again, here was a river to drool over on a perfect day with a length of water that simply screamed grayling. It ran at perfect, steady pace over clean gravel and stone, and still all we caught – or more accurately Bob caught – was a slightly battered little creature of a pound or less. The boys fished like Trojans with both fly and bait, covering the water, even skipping lunch to put in the maximum effort. It's always good to look for reasons behind success and failure equally. If a session is tough, don't give up; try hard and play with different approaches. Sometimes, though, you just have to accept a blank or at least a slow day. It's this jeopardy, after all, that keeps the challenge and adventure alive.

We had a completely different grayling experience up in Wales, on the Welsh Dee. Again, the water looked in mouth-watering shape, but this time round, in June 2022, the temperature was thirty degrees, not three! Again, John had been on the water for the preceding two days and identified three excellent grayling areas, all eminently filmable. All three locations featured those serene glides beneath rough water that grayling flock to, but this time John trialled them with corn as bait, because maggots are not allowed on the river's upper reaches. Catching a single grayling was proof enough that a shoal was resident, and then the run was kept quiet for fishing and filming.

This episode was a great success, even if we didn't locate one of the big two-pound-plus fish the river is renowned for. The grounds of the hall where we were located blossomed in the heat, and the sunsets took away our breath. Fly and bait both worked in this lush paradise, perhaps because of the warmer weather, but also because we alternated between the three swims and spread our energies more wisely and widely. Nor should we ever forget that corn is a magnificent bait, never a second best to maggots or worms.

Paul: Can I add that it was grayling that got me back into bait and coarse fishing again? I began to enjoy some really memorable sessions down on the lower Itchen (where was the film crew when you REALLY needed it?). Incidentally, I had to find another tackle shop to get maggots, as my nearest one had closed down, the victim of changing angling habits. I did find one not far away, and it was like walking into 1971. It certainly hadn't been dusted since then. Deep joy! It was a strange experience, that shop: lost in time, clinging to existence in north London.

John: I hear you on bait and coarse fishing, Paul. For four years I guided on the River Frome, close to Dorchester, deep in Thomas Hardy country. It was always very productive, with its wide variety of bends and shallows, gravel runs and small weirs. However, there was one long straight that tugged at my heart strings. We called it The Glide, and it had served me so well that I always went back there late in any given day, and often it produced a last three-pounder, back when the river was at its peak.

The surface here was smooth as glass, with barely any of the gravelly chatter that characterises much of this charming river. For quarter of a mile, the Frome simply oozed along, between five and six feet deep over an almost perfectly flat bottom, interspersed with fronds of weed. Trotting was sensationally successful, for the first years at least, but then fishing became much more difficult as the learning process kicked in. What did we do to keep the fish coming? What are the tricks that can help us all when wild fish learn to sniff out our game?

We baited, of course, with maggots for ten minutes, before sending down a float. After half a dozen trots, we'd rest the water again for ten or fifteen minutes and keep feeding as before. Often a grayling would come immediately after one of these time-outs, once their sense of security had returned. We became increasingly careful with our float retrieve up the swim. Instead of bringing the float back up the line we were trotting down, we made sure that we were reeling the float in very close to the marginal weed where it would create less disturbance. As time went by and the fish became even more spooked, we used smaller floats and slightly lighter hooks and line. We took care on the strike, too. Thinking how much commotion a heavy strike makes, so much splash and so many bubbles, we struck more guardedly, taking great care not to smash the surface as we had been doing previously. And, finally, we ditched the float and bait altogether in favour of the nymph and fly tackle, reasoning that any disturbance would be minimised still further. On 19 February 2019, this approach reaped its ultimate reward.

That morning my fellow fisherman Simon Ellis and I worked hard for little or nothing, and the decision taken over lunch was to fish The Glide for a last shot at a big one, a fish to make the sinking February sun smile. One particular piece of the run had always been just a tad better than the rest – twenty yards of flowing magic, the type of water you always tackled with a fast-beating heart, mouth dry with anticipation and mind in a state of total concentration.

Simon fished just the one fly, a red, black and orange generic nymph pattern, home tied with weight enough to sink it fast. He got into the marginal reeds, settled himself and flicked the nymph *upstream, letting it dead drift down towards and beneath him. He took his time. His casts went on the water light as the breeze. His strike indicator made no more noise than the sniffle of a water vole. This was good, very good fishing. Quiet. Confident. Meticulous. At that moment when the shadows were at their longest, the fly was taken by a fish of wisdom and experience, a fish that was fooled by the simplicity of approach and by the calm that Simon had cast over the whole session. The great grayling had been caught a couple of times in its past – which only made a third capture infinitely more unlikely – but never previously had it attained a weight of four pounds and eight ounces, a new UK record.*

We can't quite put grayling to bed without mentioning the Test, that crystalline river of dreams and plenty. Paul especially is a Test aficionado, and you'll often find him happily imbibing a little glass of white wine at the bar of Stockbridge's Greyhound Inn.

Paul and John have a dear and mutual friend in young James Buckley, keeper at Wherwell, and what this outstanding fish whisperer doesn't know about grayling isn't worth fretting about. After all, he watches these lovely creatures every day of his working life. Here are James's top Test tips.

James: I love our winter grayling days and I have strong feelings about where anglers go right and wrong. My tips start at the beginning of the day. Don't rush up to the water, silhouetting your shape, because grayling are super-quick to spook in the clear chalk-stream water. Take your time. Stand back. Spot your fish

... *The first cast is all important, but DO NOT make it a long one. Start short. Work the water slowly and extend your range. Remember that grayling shoal up in the winter, so fish each section of water as effectively as you can – and that means restraint and patience.*

Let's talk the induced take. When nothing else works, this might give you the edge. This is simply a way of tricking the grayling into rising up and taking the fly out of pure instinct rather than the usual controlled feeding behaviour. The induced take works best on fish you can see, or at least on fish you are sure of covering. Cast a nymph upstream and let it run down to the fish on a tight line. As the nymph comes within a couple of feet of the grayling, lift the rod and that brings the fly sharply up the water column, right over the fish's head. Pure instinct kicks in. The grayling doesn't stop to scrutinise the fly, it just nails it.

Smaller is often better! Grayling feed on small invertebrates like freshwater shrimp, stone fly larvae and cased caddis. Fishing a small fly right down to size 18 often produces takes that a 14- or 12-sized hook does not. Imitations like Hare's Ear, Pink Shrimp and Pheasant Tail are the ones that do the business for me. Stock your box with different weights so you can seek the fish at different depths.

Take your time. Take it slow. Enjoy the day. Search for fish. Work on your watercraft. Watch for fish rising to take emerging insects and dimpling the water's surface, even in the cold of a winter's day. Take in your surroundings, because they are glorious, and keep an eye open for the wildlife. Let yourself melt into the river valley experience and you'll begin to catch more and better fish. Look after yourself – a good hat, warm socks, tip-top boots. And look after the fish. Rest them in a net facing upstream and let them recover unstressed, however long it takes. Today's pounder is tomorrow's specimen.

PIKE, THE BIG MOTHERS

For centuries small pike have been disparaged as 'jacks' because they are generally male, squirmy creatures that can be hard to unhook. Any pike over fifteen pounds will be female because they grow bigger and become impressively gorgeous. We love them and take special care of them, because they carry extended egg-carrying ovaries as winter progresses.

John: I've fished for pike sixty years and guided for pike forty years, and I feel very attuned to them. They are mean and magnificent, and once they reach thirty pounds – or even forty – they are creatures up there with Siberian tigers, snow leopards and polar bears, such is their gobsmacking glory. I've been entranced with them since I could walk, almost, these water wolves of childhood dreams.

Paul: When we were young our favourite pike method was twitched sprat, a modification of sink and draw. We'd let the sprat lie for up to fifteen minutes, and then stir it into life with a turn of the reel and lift of the rod tip. The hope was that the pike would think it was a sick or injured fish and have it. By the way, John, I have had RIVER pike of twenty-six pounds and FLY CAUGHT pike of twenty-one pounds, so I've not done badly.

With pike, it's all about windows of activity. By that we mean those special periods when pike come on the feed after hours, days or even weeks of inactivity. Pike do not feed for prolonged periods like many fish do. They are meat eaters, not grazers, and a big meal can take a long while to digest before hunger kicks in again. Pike waters seem almost fishless. There might be a bit of 'jack action' to keep you amused, but the big girls will lie still as a log, almost comatose until hunger rouses them, and nothing any angler can do will hurry them. What is imperative is to fish as well as you can so that you are ready for the moments when the lake switches on and your fishing life can change in an instant.

Understanding the rhythms pike follow is largely beyond us. We like to fish on dark, wet, windy days of low barometric pressure, but we have had sensational days when the pressure has been high, the sun has been bright, there has been a frost and you think you don't have a pike in hell's chance of a catch. It's also true that many waters fish well at first and/or last light, and if you can arrive early and stay late, do so. However, gravel pits especially do not always conform tradition. On many, the hours between 10 a.m. and 2 p.m. can prove consistently the best, but never forget, a monster at dusk is always an exciting possibility.

Apart from otters, the only creatures big pike need worry about are anglers. And you *must* learn to be happy around pike before you fish for them. Not all pike that are caught go back well, and pike that go back well and unharmed learn from the experience. The key is to look for your own pike hotspots. Don't go with the crowd. As we're writing this, John is preparing to go on a recce to the far ends of Scotland, to a pike lake whispered to hold monsters, but this is a high-stakes adventure with little solid fact to go on. Perhaps the search will be dogged by failures and disappointments, but that is integral to the pike fisher's game. Mystery. Jeopardy. These are the magic ingredients of a winter pike challenge.

Most piking is done in the cold weather, beginning in October and finishing in March, but wherever and whenever you pike fish, remember that they are no fools. There is an ancient fallacy that predators are savage and headstrong, somehow blind to danger. They are not. In our experience, big pike are as cutely clever as big carp, chub, trout or any other species we fish for with utmost caution. If a big pike has been caught on any method or bait, then she is highly likely to avoid such approaches for the rest of her life. Her memory will not fade, and she will see out her days uncaught unless an angler thinks of pursuing her with ideas that are completely fresh. It is not unusual for very big pike to vanish for years, despite heavy fishing pressure. Just when everyone thinks she has passed, she astounds by being caught again.

So, treat pike with respect. Don't tramp to the bank. Think about your rig. Don't fish too heavy and avoid any unnecessary weight on the line, because a cautious pike will feel it. If there's a hint of resistance a wary pike can eject a bait as fast as any fish swimming. Try different baits: the wider your selection the better. Work a selection of lures at different speeds and at different depths. Sit right on your rods and look for even the slightest indication of a pike pick-up, and never expect line to rattle out till you strike. Sometimes the slightest tightening – or slackening – is all you'll see.

Don't spend the day asleep in a bivvy. Keep alert, keep looking at the water for any signs of feeding fish. If you see a pike roll, sometimes just the head breaking surface, then get a bait on her fast. Look for big areas of calm water in the midst of ripples. We call these 'flats', and they are indications that a serious fish has turned under the surface or is feeding off the bottom and sending up the gasses trapped there. If you see these 'flats' it could well be that your bait has been nudged or picked up, even, and that the oils it contains are being released. These oils are floating up through the water column to the surface, where they will spread like a slick, a sure giveaway of pike action beneath. Large patches of bubbles have to be investigated. These are created by a pike rising from the bottom silt as it awakes and begins to look for food, or even by a pike hitting the bed in its pursuit of an eel or roach hiding in the bottom detritus. Miss these signs and you might miss the fish of a lifetime.

But to return to the most vital concern of all, KEEP YOUR PIKE SAFE! Strike fast at any sign of a take. Have the right tools but, vitally, know how to use them. Always start piking with an experienced friend or have unhooking lessons from an angler or a club. We can't advise you better than to join the Pike Anglers' Club of Great Britain, or at least go on their website, where they describe in perfect detail how to unhook these fish with your safety and that of the fish in mind. Blood on the banks is not what any of us want to see. Do things right and there won't be a single drop.

Gone Fishing went to Lough Erne in Northern Ireland in November 2018, and you'd think success had to be guaranteed. Erne is a vast waterscape – just the lower basin is eight miles long and five miles wide, and it is festooned with countless islands and boasts depths of 200 feet or more. We were based at the excellent Watermill Lodge and guided and mentored by the legendary chef Pascal Brissaud. All in all, spirits and hopes were as high as the Irish sky. But what happened?

Not much, to be honest. For three windswept, storm-tossed days Paul and Bob really worked that water with every conceivable and legal method. They never stopped as Pascal urged them on. The boat fished the open water, the deep water, the shallow water, the bays, in the wind, in the lee, around the islands, over rocks, in the reed beds, new water, favourite water, just about every scrap of water you could think of and then some. Every lure, every bait was tried at every depth and every speed. Yet! In all, three jack pike were fooled and one got off close to the boat. It was a paltry result for so much effort – good effort, at that.

The time of year wasn't bad, nor the weather. Pascal knew every contour, every creek, and he ensured every cast the boys made was a good one, a cast with a chance. So what went wrong? What are the piking lessons we can all draw from this?

Firstly, pursue wild fish and you never know what the outcome will be. On natural waters jeopardy is a massive part of the game. You might have every box ticked – or at least you think you do – but failure still triumphs. Second, even a vast water like Erne is affected by angling pressure. Those educated pike know the sound of a boat, the splash of a lure – they know to be wary. Even with hindsight, neither of us can think of much done wrong, of any obvious errors made. It was simply a long way to go for six pounds of pike.

But compare that Irish adventure with an East Anglian trip undertaken by us, one snow-coated day in February 2018 – before the days of *Gone Fishing*, but with Bob as our guest. We were on a large pit that John knows well, and despite the vile weather and the extreme difficulty of extracting the boys from their hotel, he felt confident as a result.

The morning wore on, with little sign of life; even the birds were quiet in the gloomy cold of a stark mid-winter day. Remember, though, all our previous talk about windows of activity? A little after 2 p.m. there was a softening in the wind and the temperature rose two degrees. A sprinkling of surface roach activity was sparked, and almost immediately Bob's deadbait was taken. That fish, Bob's first pike, weighed an extraordinary twenty-eight pounds and twelve ounces, and if you bump into Bob bankside in years to come, he will claim it a 'thirty'. It was a life-changing result in angling terms – and surely an example of success on a water where you have faith and experience.

But how did Bob manage to nab a 'thirty' (supposedly)? On any pit there will almost certainly be a steep drop-off close to the bank, and pike love to patrol this, providing there is little disturbance (it's a given that we won't make a commotion, but do everything with caution – ensure rods, bags, boots and anything else are placed with care).

The first step is to look at the pit's geographical layout. As most winds blow from the south-west, a good area to focus on is the bank to the north-east. It's around here that prey fish are likely to congregate, and if you find the shoals you are halfway to cracking the location of the predators. We've written a lot about wind direction, but we shouldn't ignore other sections of the bank. Our pit – in fact, let's call it Bob's Pit out of love and respect – is no more than six hundred yards long. A pike can drift from one end to the other in an hour or less when it is actively on the prowl. It's the same with depths. We always stress about finding deep water in cold weather and shallows in the spring, but a pike can have breakfast in ten feet of water, a mid-morning snack at six feet and lunch almost off the surface.

Our advice is not to get too wound up about these details, but rather look for 'structure', as the American bass experts say. Structure is a word for features. On our pit we are looking for overhanging alders and willows, especially where a branch has broken off and lies waterlogged, forming a snug protective ceiling for a resting pike. The bank, with the endlessly thick reed beds, is a magnet for silvers that spend their lives being hunted, so it's where many pike will like to lie as a result. There's a bridge near Bob's Pit, too (where Paul fly fished in one episode of the show), and the pilings are a favourite ambush point for predators. A stream runs into the lake there, leading from a ditch that's full of three-inch roach. We liken it to a fast-food diner, and we always see half a dozen good fish hanging around the serving hatch.

Simplicity is our watchword, and we keep rigs and approaches as basic as we know how. We spend a lot of time with baits close in, and we know we can fish lighter than most pike lore would have it. Ten- rather than fifteen-pound line can make a lot of difference. We'd never endanger any pike by fishing too light, and we have never lost a fish because we are not playing them over gravel bars or through distant weed beds. Another advantage is that we can use smaller hooks, 8s and even 10s rather than 6s and 4s. We even favour larger single hooks and sometimes circle hooks. We don't need heavy leads to achieve distance, and a couple of the large split shot known as SSGs will suffice for a ten-yard underarm flick out. We'll even freeline a deadbait at times for absolute minimum resistance. At this range we are right on the rods, looking for any sign of a pick-up, however careful it might be. Everything is close in, white-knuckle stuff that turns piking into a very intense experience.

This is as much natural history as it is fishing. At Bob's Pit there has been heavy cormorant predation for many years. The eventual result is that silver fish of more than half a pound become very rare and the average size is two to four inches, but there are lots of them. This has meant that over time the pike have become used to feeding off more smaller fish rather than satisfying themselves with just one or two big ones.

Our light approaches are perfectly suited to these smaller baits and, of course, many small food items mean the pike have to feed for longer periods and more frequently than in the days when a two-pound roach or perch sated them for a week or more. We've even created our own feeding patterns by pre-baiting with chopped deadbaits; introducing a few kilos of chunked fish every week can wean pike onto taking two-inch square pieces of sardine, mackerel or roach in preference to anything else. You can fish these small fish baits under a waggler float, rather like you would for tench, but with a wire trace, of course. How We Fish, you see, is often unexpected, but it's always interesting, engaging, simple and fun.

Paul: Blooming pike! Wish they'd learn to speak English.

ROACH, THE GENTLE GIANTS

It was right at the start of the winter of 2022, and the *Gone Fishing* team were on the Hampshire Avon, close to Salisbury. It was the second day of filming, and John had identified a glorious slow bend with a steady current on the outside and a large eddy on our near bank. In the morning session, the boys had trotted corn and maggots for a stream of quality dace.

It was classic *Crabtree*, a grey-brown day, a hint of rain from scudding clouds. The willows were bare, the temperature was rising as dusk approached, and John and Paul had that feeling that something momentous might just happen. The cameramen looked at their meters, and in the last of the light Bob and Paul moved back onto the bend. The second time, the floats were pushed well up the line so that the baits were fished hard on the river bed and completely static (see illustration below). For an hour before their arrival, John had dribbled corn into the slack, and by 3.30 p.m. the stage was nicely set.

Bob missed a bite and caught a chub, but it is Paul's story that interests us most: his float cocked after some fifteen minutes and sailed away. The strike met something sizeable, a fish that turned immediately on the surface in a blaze of gold and a flash of crimson fin. Roach! Then it was off. A simple hook-slip, no way the fault of the maestro. We all felt the old sickness in the stomach, that moment when the world stops turning. Not only was it a roach, it was a big one. A two-pounder, probably. This is the fish that so many anglers regard as the holy grail, the pinnacle of achievement. Gone. It's a loss both of us will live with forever. (*Paul: I will add that I DID land a ten-ounce roach that day caught on float with the bait dragging bottom. It might not have meant a lot to John, but I loved it.*)

All roach are creatures of wonder. Our careers started with these pearly-scaled beauties, and we'd be happy to fish for

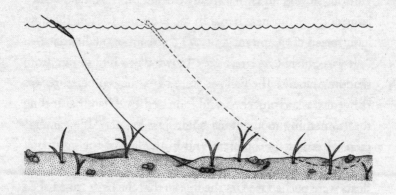

Stret-pegging for a roach

194

nothing else, such is their charisma, charm and challenge. Of course, you can catch roach from still waters and in the summer and autumn, but traditionally they are at their best in the cold-weather rivers. That's when their condition is at its finest, when the silver of their flanks hints a steely blue and when trotting the current is the champion of methods.

Broadlands on the lower Test is the place to go in the winter if you want to see river roach trotting at its most refined. Once the razzmatazz of the trout season finishes, a syndicate of coarse rods takes over the river in the winter, and these are anglers of the highest accomplishment – experts in the field of float and centre pin. It is deep, fast, unruly water that burbles along over gravel and dark holes. There's often a wind on the water and control has to be perfect if a big roach is to be deceived.

However, there are many roach rivers where fishing a float is less of a task and the water isn't one of such dazzling and baffling complexity. The Wensum is one such river – where Paul has fished and which John knows like the hairs on his hand.

When John guides on the Wensum or when we fish together, there's a particularly lovely stretch of middle river that we always head for. We'll stop at the top mill pool; it's very pretty and often home to decent roach aplenty. Then we'll move from there and investigate the half-mile downstream, which is quick, gravel bottomed and shaded by endless willows and alders. Finally, we'll explore the deeper, slower stretch a further three miles downriver, where the Wensum is paused by her next mill in the chain. All three locations hold great roach at times but ask different questions when it comes to the actual fishing for them.

There is an ecological issue here. The Wensum, like many lowland rivers, has a whole string of mills along her course. The Environment Agency (*Paul: We've done pretty well to avoid controversy until now, I guess*) would remove these mills if it could. The Agency has its own reasons for wanting to do this, but as anglers we would counter that the mills have been here a thousand years and the fish populations have been prolific for most of that time. Most of the problems that the Wensum, and lowland rivers like her, face have happened in the last forty years and cannot be blamed on the Saxon or Norman millers. The present structure of the river also allows the roach to find different water types and conditions as the seasons turn. So, from March well into June, roach will be in the fast, shallow water close to the mill, where they will spawn. From June to September, they might drop downriver to feed hard on rich insect life found in the gravels, and then, when frosts set in and the water cools, many roach drift to the deeps where the winter can be spent in comparative comfort.

So, March in the mill pool itself can be glorious. A thirteen-foot float rod, reel with four-pound line; hooks, shot, four pints of maggots, a net, and the day is yours. We choose a swim around five-feet deep, with steady flow over a clean bottom. 'Steady' is important. We don't want to see big boils that make the surface churn like a witch's cauldron. Placid is what we are looking for – a fifteen-yard run that can be tackled with a 3 or 4BB stick float, size 16 hook and double maggot bait. We like the shot placed just beneath midway between float and hook, say at three feet, six inches if the swim is five feet deep.

We take time getting the depths right because roach like the bait just tripping bottom. A few exploratory casts will give you a good idea of the float setting. Most fisher folk use far too little bait, whereas we put a generous fistful of free offerings in every single cast. You've got to draw fish to you and then get them feeding, after all. It can take you an hour and cost you two pints of maggots, but then the magic begins. These are short trots in water of easy pace, so control isn't too hard – nothing like the turmoil of water facing those poor souls on the Test. Keep at it, concentrate, and if roach are there, they'll oblige in the end.

A few further thoughts: a hooked roach is very vulnerable to pike attack, and if there are too many predators around, then sorry, you'll just have to move and fish elsewhere. Not getting bites? Try resting the swim for ten minutes. Keep feeding, but by not fishing you are building up confidence. Try 'holding back'. Simply check the progress of the float so that the bait rises in the water column, often more than a roach can resist – it's just like the induced take that our mate James Buckley described when nymph fishing for grayling.

Remember that if you catch roach of two ounces or two pounds you are the luckiest roach fisher around. The size does not matter. These are stunning fish, coming in a glorious stream of braided silver and scarlet. Watch the kingfisher, feed the robin, share the swim with your best pal and you'll reminisce about the session for years. Relish the odd gudgeon or dace that comes along and that bristling perch. You might have a tussle with a chub, be broken by a barbel or land a stray bream. This, our friends, is simple, wonderful fishing. We purr with pleasure just thinking about it.

November is a fine time to be float fishing for roach in the quick water below the mill. There can be big numbers massing there in milder late-autumn days as the shoals wait to head to the downriver deeps. Fishing is much the same as in the mill pool itself, but the water is that bit shallower and more speedy, so you will have to work harder than before. Consider, too, that your float might have to travel twenty or even thirty yards to find fish, and new issues of control come into play. But fret not. It's all the same game, in truth, and the basics are the same. Practice will make you perfect, or at least as good as you need to be.

John: Three years ago I was blessed enough to see a pod of big roach on this very stretch, and what a sight of wonder that was. These were fish so graceful and so enormous that fishing for them was an hour of epic, excruciating ecstasy with a fair dash of horror thrown into the mix. My mates Pingers, Ratters and I made up the gang of three that day, all crouched beside a five-foot run over gravel, between willows, and still threaded with remnants of summer weed. The day was bright enough to see the shapes of the seven fish as they drifted in and out of the willow roots, exploring open water and even digging in the bed for caddis. The baiting began, a steady feed of white maggots that dribbled along the twenty-yard stretch of river like snowflakes in the winter sky. Little by little, with no suggestion of haste, the roach began to take a smidgeon of interest. An hour passed before the maggots were intercepted with an intensity that made actual fishing a viable proposition. A single rod was set up with four-pound mainline, three-pound hook length and a size 16 microbarbed hook. We chose a stick float built to carry three BB shot, nipped one shot a foot from the hook and placed two more closer to the float. It was just enough weight to get the double white-maggot bait into the zone where the roach were feeding. Pingers took first cast and we watched in breathless silence. Two yards, five yards, ten yards and then the float buried, the rod lifted and an imperious roach broke surface.

It weighed two pounds, two ounces, a stunner, a prize beyond compare. And away it went, the swim quiet again. The morning was young and we knew patient feeding might just bring those fish back again – so we settled to the task. First one roach, then three roach, and after forty minutes all remaining six roach were back in open water. My cast was instantly met with the smallest roach of the pack, around a pound and a half, and this was hustled out so fast that the big five barely noticed its absence. Now it was all down to Ratters. Ten minutes of feeding and the biggest roach of all sipped in the size 16 and surfaced, balletic, beautiful, intent on escape. And it was off in tantalising seconds, like Paul's Avon monster, another damned hook slip and the game was done. Agony and ecstasy like nothing else in life.

Three miles beneath this battleground, the Wensum slows, deepens and widens out, held back by the next mill in the chain. John has a history on this daunting and featureless length of river, but over the years long-term baiting with bread mash has brought fish to his net in an imposing number: eight three-pound roach and scores of big 'twos' fell here in the Wensum's heyday. Big roach do feed in daylight on stretches like this, but most come at dusk and well into dark; by mid-winter many of the biggest roach have made their way to their cold-weather quarters.

Night sees the roach angler as ambusher. A campaign like this calls for dedication over two or three nights, luring the roach into the area, weaning them onto the wetted, mashed bread thrown out onto the river bed. Now it is time for the quivertip or even a small ten-foot bomb rod, four-pound line straight through to a size 10 hook and bread flake. Two SSG shot should be weight enough to hold just off the main flow, exactly where the deep central channel begins. The fishing might be slow, results even non-existent, but these are special sessions, filled with the most exquisite blood-pulsing excitement. A single bite might be enough, one fish to top a roach career.

Indeed, fishing for roach takes us into a kind of dreamland, and when a roach rolls upriver just after dark you know from the solemnity of the sound that it is huge. An owl calls and is answered, perhaps by its mate, from the trees on the island. There are whispers of wind in the rushes that ruffle the quivertip and make the heart skip a beat. The evening ticks on by, a cast made every twenty minutes, the time it takes for the flake to soften and fall from the hook. Stars sprinkle the heavens, and if a big moon is up you can watch the tip by its light and switch off the torch; your night vision grows and grows until you can see the owl, even the vole it is hunting. You are at one with the roach, the river, the whole flood plain, and all the silvery magic Jack Frost lays out before you if the night turns cold.

This is fishing of a different sort, of a new intensity, and though the chances are that you will trudge off home without a bite, you really hardly care. You even want it that way, because you know that the purest prizes in fishing come after supreme effort. Wet and cold? What do they matter on a session like this? These are the adventures fishing can bestow, even in modern, over-urbanised Britain. No sport has dimensions as resonating as ours, a passion for all your years.

CHAPTER TWENTY

SPRING SESSIONS

Paul: 'April is the cruellest month' is the opening line of T. S. Eliot's The Waste Land. *He obviously wasn't a fly fisher, then, because 1 April is a traditional trout opening date. Perhaps he could have been a coarse fisher bemoaning the river closed season, but I doubt it. Could have been he was just talking about the weather.*

We all know spring weather varies between wonderful and dire, but on the bright side (*Paul: Where we make a habit of looking*) lengthening evenings, warmer days, emerging weed, reawakening insect life, fish feeding harder and longer are an angler's dream. When you get to our age that feeling of light and life is a rebirth, as though we're emerging from a dreary cocoon following a wet, deadly winter. March gives way to April and May, and by then we are luxuriating in a world of colour, not monochrome – marvelling at the golden glow of the waterside willows coming into leaf.

We're no longer rugged up like the Michelin man. We can actually move again, on the hunt for trout or tench, listening to the song of the nesting birds, watching the swarm of the toads now that Jack Frost is gone for another year. That first cuckoo call. The first picnic that's enjoyed, not endured; and the prospect of a drink in the pub garden when the fishing day is done.

The harsh face of spring sees easterly and northerly winds set in for weeks, blowing in their misery from the Arctic, from Siberia, from polar-bear land. There's ice on the windscreen come morning; fly life shuts down, trout don't rise, tench don't bubble and there's no sign of the surface-skimming swifts. Snow in April, sleet in May and you're back, ferreting for your winter thermals.

John: It's only two years back that Paul and I were on the upper Wye in just such conditions. The Wye was as dour as a ditch, and though there were glimpses of sun, not an insect appeared, not a trout broke surface. When Paul fishes, he does so hard and he covered the riffles, glides and pools supremely well, in my eyes. He fished traditional wet flies in teams of two and three, down and across, moving steadily down the run. Two hours' faultless fishing for nary a touch. It was simply one of those days you have to accept and abide by. Acceptance of days like this comes from a good fishing education, which his dad gave him. Wild fish are always that little bit ahead of you and always have the final word on tough spring days like these.

There is good spring weather and bad, and our tip is to keep your powder dry for the latter and wait for the former to fish hard. Supposing the wind is blowing from the south or west, fish where the water is exposed to its effects because there'll be oxygen there, significant undertows and food items stirred up into the water column. Watch where the sun hits the water and try there if you can. Emerging weed growth is a real fish magnet, and put in time around the shallows where fish will be looking to spawn when the water warms sufficiently.

Increase your baiting because fish coming out of winter are hungry and coarse fish will be feeding up prior to spawning. Also, remember that the number of natural foodstuffs is multiplying exponentially and you are competing with a myriad of insects for the attention of the fish. It's no good putting out half a can of sweetcorn and expecting to hold a ravenous pod of tench; spring is one of those times that you reap what you sow. Fly fishing? This is the time that bigger flies and lures will be at their most dangerous.

We cannot talk of 'bigger' flies and springtime without reference to mayflies, the exquisite creatures that rise in spiralling flight from the water, their wings of gauze aglow in the sunlight. There's not an exact date for the mayfly hatch, but on many southern rivers late May often sees its climax and they become the insects of the late spring. While rivers across the country experience the mayfly phenomenon, it is the valleys of the southern chalk streams where they are at their finest.

There are times in the year, fewer in frequency now, when the flies fill the Test valley like a snowstorm, nourishing the vast array of lives that depend upon them. Trout gorge, throwing themselves into the air to take the succulent insects in flight. Swans feast, scooping spent flies from the surface of the water and dipping their graceful necks beneath to intercept the rising nymphs. Hobbies, those tiny falcons, rise and stoop for them. Eels snake from the river's bed to gulp down the nymphs and emerging duns. Bats assume the aerial attack from the fall of dusk, and even the barn owl hunts the strays as he quarters the water meadows in the last of the light, the end of a day of plenty.

One blessing is that silly o'clock alarm calls aren't the necessary evil they are come summer proper. Cold nights linger on and the temperatures that climb with the sun will see all species feed best from mid-morning and well through the afternoon. A mild, still May evening can herald one of your best sessions of the year, whatever your target, but most especially for tench, unquestionably the coarse fish of the spring. If the nation has a favourite fish then, according to polls taken over the years, tench are always in top spot or close behind. Look at them. Their smooth, polished flanks, their coal-black fins and red eye startlingly set against the head of deep mahogany. Their balletic poise in the water, their ability to hang with such resplendent ease, their serenity and yet their arm-wrenching power when hooked. What a package of April delights.

TINCA TINCA – OUR GLORIOUS TENCH

We've had bad times and good tench fishing over a lifetime of springs. We've come to see tench as living barometers, quiveringly alive to all the vagaries of spring weather patterns.

A shift in the direction of the breeze, a shaft of sunlight, a rise in air temperature and a faltering hatch of alder flies can trigger a tench bite, seemingly out of nowhere. In spring, parameters change in an instant and a sleeping tench becomes a feeding one. We might have been failing for hours but the fault has not been ours, and now Nature has flicked her switch and all we need to do is capitalise. A bleak day becomes a red letter one, a miracle that takes only seconds to perform. In spring we can do the right things but always accept that our tench obey a higher power, and that when conditions are wrong every one of our best efforts can be doomed to failure.

We've talked about windows of opportunity elsewhere in the book, and nowhere are these more relevant than now. Many of our dreams begin and end with tench, but this fish is capricious beyond almost any other species. For us, this is almost all to do with wind direction. A wind from the south or west is excellent. A wind from the north is bad, but a wind from the east is disastrous. Simplistic? We don't think you'll ever prove us wrong.

Much has changed in the tench world since our respective childhoods. For the past quarter of a century there has been no closed season on still waters, and we have therefore learned to fish for tench in April and May, perhaps the best months of all, when the waters are waking and the tench are feeding hard. Tench waters have also seen a shift in focus away from estate lakes towards more recently dug gravel pits. Smaller waters have been replaced in part by larger ones, and in these past years feeder fishing for tench has become more common than using the traditional float.

Tench baits have also progressed radically. When we started our tench adventures, baits were always bread flake, maggots or lobworm. Then in the seventies sweetcorn rained into every tench lake in the land. Now pellets and boilies are not just seen as carp baits but essential for tench, too. All this time tench have also been growing larger. In the 1950s a record tench was seven pounds; today it is more than double that weight. This century a specimen tench is eight pounds, but when we were boys a 'three' was a monster to be proud of. None of this has changed our passion for tench or taken away any of the glory that surrounds them like a halo. They look gorgeous, they are fascinating to catch and nothing can detract from their desirability.

Kingfisher is the famed pit in mid-Norfolk that we know best and love most. It is home to famously large and beautiful tench. The lessons we have learned there have translated to pits all around the country, and wherever you make your start you can have confidence. All the pits were dug with a marginal shelf, and the bottom shelves deeply and steeply just a few yards out. Tench adore this feature and will follow the ledge when they look for food. They travel in pods of a handful of them in less-populated waters, but in their scores at Kingfisher. These pods are territorial and cover a few hundred yards in range at most, but consistent baiting does concentrate the tench to visit certain tight areas more than others. Also, if you put in a lot of bait over a sustained period, you will draw the pods together into a large shoal of fish, and that is when sessions have become almost historically special.

You have to remember that if conditions are good then tench feed ravenously. You cannot put in too much bait: a can of corn or a cupful of pellets goes nowhere and is nowhere near enough to persuade a wandering group of fish to drop down between eight and a dozen feet to the bed and feed. The best results depend on heavy feeding, either on the day or preferably over time, if you can get to the water frequently. Our advice is to put bait in the evening before fishing if you can, although we know that is difficult on a public water where a swim can easily be taken.

An excellent tench recipe begins with a cereal-type base, like the dog food Vitalin. Mix it with water until it becomes firm, not crumbly and introduce plenty of your hook bait samples, corn, pellets, boilies, maggots or even all of them, if you can. Five kilos of bait should be enough; scoop it out to the bottom of that first ledge, generally around five yards from the bank.

Close-range fishing like this is perfect for a float approach, but there are many occasions when tench shoals favour similar ledges further out into the lake. These are harder to find and far more difficult to bait up; you'll need a catapult to get balls of bait out, which is a long process. You will also find fishing a float at distance far more difficult and feeder fishing becomes something we have to do. Results, though, can be phenomenal.

There is a special swim at Kingfisher on the island that has been sensational over the years. A gravel bar runs forty yards from the bank and it's like a highway that shoals of tench from all over the lake like to follow. Sessions begin with us baiting hard, ball after ball catapulted out, landing just short of the bar, in about seven feet of water. We don't stop, sometimes not for an hour, and we'll put in more bait throughout the day. We'll use either a method feeder with a boilie or a blockend feeder with maggots and set them up with eight-pound line and size 12 hooks.

These are pretty much self-hooking rigs, and bites are spectacular, the rod tips just hooping round. It's non-stop fishing and it's not always straightforward. Sometimes natural maggots will be refused and we have to put plastic ones on the hook. These look odd, but they are buoyant, counteract the weight of the hook and get sucked more easily into the mouth. There are times when we must dredge the bed with nets and search for caddis grubs in their cases made of sand and gravel pieces. Once shelled, these grubs are yellow or creamy white, and we put three on a size 14 hook and the tench just adore them. This is not the simple fishing we prefer, but there are times when modern baits and techniques are necessary and this is one of them.

If it's pure tench fishing you want to hear about, then a special session with Paul and Bob has to go down as an epic. This time we're on an estate lake, but once again the drop-off worked wonders, close in where the margins shelved to six feet deep some ten yards out. Baiting was vital, but now it was simple and direct, just a kilo of boilies laid out the evening before fishing. And glory be, there was not a feeder in sight, just glorious, hand-crafted waggler floats. Nor could the rig have been made more simple. The floats were fixed and cocked by locking shot and set at seven to eight feet, nicely a foot or two over depth. There was no weight on the line and this fell loosely from the float to the bait, a boilie, on the lake bed. (Again, Part Three: The Nuts and Bolts of What We Do will highlight this most basic but efficient of rigs.) Tench adore red above all other colours, and 12 ml baits on a short hair and a size 10 hook completed the entire set-up.

In practice, this is so easy it's laughable and so direct as to be devastatingly efficient. Day long, as the cuckoo called, those floats just vanished in a way not far short of magic. These were big tench caught in a heavenly place and in near-classic manner. When sessions go this perfectly, you walk away on a cloud, there's nothing like it in sport. You feel like the best angler there's ever been, that you are invincible. And then, next time out, it all comes tumbling down, the whole tench fantasy you have constructed around yourself. Just like at Lost Lake, John's nemesis.

In terms of looks, our secret lake, appropriately named Lost Lake as it is so hidden in Aldercarr, has to be a jewel in the crown of all tench waters. Six acres nestle in wood and wetland, not too deep, weedy but not overly so. A haven of bird life, deer and otter, Lost Lake is a pool of absorbing serenity, a world apart. The water is crystalline, and once the sun is up above the tree line we've always been able to see the tench, happy in their paradise. They are very often hungry, too, and we occasionally put in a bucket of bait and watch the great dark shapes congregate, tip and feed with volcanic swathes of erupting bubbles. The lake is fished only by us and completely off any grapevine, so the scene is set for great tench adventures. Which never happen. Though they feed, those tench very rarely pick up a bait on a hook, so any capture is red letter stuff. If we list all the ruses we have tried over the seasons at Lost Lake, you'll see how far ahead of us those tench are.

We start with the obvious. Baits. Boilies of all flavours, colours and sizes. Maggots, corn, casters, both real and plastic. Baits hair rigged and side-hooked. Basic baits like flake, slugs and lobworms. Intricate baits: Ratters threaded six plastic bloodworm imitations onto a hair and had one big tench and one tench alone. We've fished feeder and float, a float sometimes as small as a matchstick. We've fished dusk, dawn and through the night. We've raked swims in classic tench fashion. We've fished frighteningly light, in close and far out. Still, we have almost always failed and have not an inkling as to why.

Lost Lake is the finest example of how tench waters differ hugely in difficulty from the easy to the nigh on impossible. The more challenging waters generally are naturally food-rich, or have few fish or are heavily pressured. But Lost Lake holds high numbers of tench that are never fished for and feed greedily on what is offered to them. This makes little sense, but there's some comfort in that: wild-bred fish are so challenging because their survival senses are uniquely tuned. This takes us back to the cornerstone of our fishing philosophy – that fishing is very much like exploring natural history, and in doing so we become fishing detectives.

It's nearly time to leave our springtime still waters, but how can we, without some mention of bream? Bream and tench are found together on many still waters, where the two species

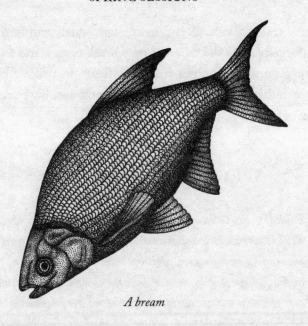

A bream

intermingle and get caught often cheek by jowl in the same
session. But this doesn't always happen, and bream do merit
a quick discussion all to themselves. Bream don't generally
like the margins of any lake, and you'll find them further out
in gullies and troughs, where they spend the majority of their
lives. On mild, breezy days you'll see them breaking surface,
often at range, and that's where our binoculars serve a good
purpose. Once you've found a bream hole, remember it well,
because the fish can stay in there for weeks and they help you
out on days when the tench prove tough. Bait big with catapult, and use the same rigs and baits as you would for tench.
There are those (*Paul: Aren't there, Bob?*) who have no feeling
for bream, but not us. History has it that they fight like wet
sacks, but not all of them – when they get big they can give

you a real run-around. Bream are both stately and imposing, unforgettable in the moment they break surface and a spring sun gleams along their immense flank. Just like Achilles' shield, they look all gold and imperious, and we personally can't get enough of them.

SPRING SALMON SESSIONS

Paul: If you have an opinion about salmon fishing you're opening a can of worms, because somebody, somewhere will have a different opinion and disagree, probably vociferously. But I don't care and we've got to start somewhere. One thing I think we can be sure of is that salmon don't feed when they enter freshwater – cue howls of derision from those who say they've seen salmon 'take flies like trout' and 'if that's the case why do salmon swallow worms?' Be that as it may, if salmon were to really feed when they entered the rivers of their birth on their journey to spawn, they would wipe out their own population. They are, after all, voracious predators. Nature's solution to self-inflicted extinction is to suspend the salmon's desire to feed in freshwater. So, how on earth can they be caught? A good question, and most of the time they can't. But as anglers we rely on the feeding instinct or memory, aggression, or irritation with an intruder, in this case our fly, lure or bait.

The salmon season in the UK begins in some rivers as early as January or February and is called spring fishing. As any child knows, January and February are winter months, not spring ones. And March and April in Scotland can compete with some

seriously cold days in January. Such days have their own appeal
– the wee dram (it's traditional to escape to the hut for coffee and
a nip of something stronger, though not for me these days) and
the fact that it's dark by 4.30 so you can get the hell out of there.
These are things that give you hope when you make your heroic
approach to the river with ice in the margins, ice forming in the
rod rings and the occasional blizzard roaring through. Not for
the faint-hearted.

I remember such a day on the Findhorn, one of Scotland's, if
not the world's, prettiest rivers. It's a smallish- to medium-sized
river and I've fished it below a serious gorge. Crunching across
the frosty undergrowth and over slippery rocks, with extraordinary red sandstone cliffs on the opposite bank and with small trees
and scrub growing in seemingly impossible places, it's a magical
place to fish. It has a good 'spring' run, or as good as can be
expected these days, and although small by comparison with
rivers like the Tay or the Tweed, the approach is fairly standard
for the early months. Spring fish are ready takers, we are led to
believe, but don't let that fool you for one second. Be prepared for
blank days! You will have them. John and I have had many. But
they are character building (ha!), and remember, 4.30 p.m. is not
far off (and you don't need to start fishing in such cold conditions
much before 9.30 … ish).

If you're fly fishing it's a good opportunity to practise your
technique. Spey casting is what we use to get a fly – a big salmon
fly, mostly – out long distances. You'll want a long rod, perhaps
thirteen or even more feet long. We call this a two-hander, as it's
too heavy to hold in one hand – unless you are Desperate Dan or
Bob Mortimer, of course. The technique is tricky but easily picked

up if you get some lessons from a professional. If you can't Spey cast I suggest you get some lessons before you do anything else. It's a joy in itself to see your line unfurl gracefully and land like thistledown exactly where you want it, despite the howling gale blowing directly into your face from the Urals. Needless to say, this doesn't happen very often! But persevere. I don't get to fish for salmon as often as I'd like, so I take a while to get back into the rhythm. I've read too many books with complicated diagrams purporting to teach you how to Spey cast, but unfortunately you can only understand them once you've had lessons and practice. Of course, you don't have to Spey cast, you can do the more conventional overhead cast – or begin with tuition, if you want to master both.

The standard approach and time-honoured method in the early part of the season is to fish a large fly deep and slow; if you're spinning, it's the same principle, try to keep your lure deep and slow. Cast your fly slightly downstream and across and put a big upstream mend in your line (that is, flick your rod tip upstream as your line lands on the surface) to encourage your fly to sink. Consider technology – fly lines in particular have been transformed in recent years, and there are many specialist profiles and densities to help search the layers of the water. In the past, a medium- or fastish-sinking line was always the way to go. Nowadays, there are lines that are much easier to cast than a full sinker and come with an array of tips of different densities to help you explore the depths. When fishing a sinking line or sink-ing tip remember not to use a long leader. You want to get your fly down, and a short length of monofilament or fluorocarbon helps keep your lure in line with the sink tip.

Some people like to have a heavy fly, like a copper or brass tube, or a conehead, which can be constructed from tungsten, if you want to truly dredge the river bed. Others prefer a lighter fly fished deep. As a general rule in the spring, we reach for a fairly big fly, two to three inches and often with orange, yellow and black as the go-to mixture of colours. It's why the Willie Gunn and Gold Willie Gunn are such enduring flies. Another popular choice early season is anything in yellow and black, but there are many salmon fishers who don't think colour is that important, size profile and action being much more crucial ... very much like they are on the dancefloor, though who would truly rule out a dash of colour there? Certainly, fly variety makes for a more interesting day and it often pays to change your approach.

So, out goes your perfect cast, your conehead Gold Willie Gunn turns over perfectly. A quick upstream mend and your fly is fishing well on a ten-foot fast-sink poly leader with three feet of twenty-pound-breaking-strain fluorocarbon. It swings round beautifully in the current, and as it enters the slack water you handline in a few yards before taking a couple of good steps downstream, disentangle the line from your wading stick and cast again. If the water is very fast you might want to cast, put a mend in the line and then take a step so you give the fly more opportunity to sink. On a crisp spring day with a bit of sun and a gentle breeze, you won't mind that you haven't caught anything after three hours.

My first spring salmon was caught on a brown and gold Devon Minnow spinning on the Dee in Scotland, when spinning was allowed there. These days it's fly only, and with the advances made in modern tackle and casting techniques I don't have a problem with that. I'm not against spinning, though. I just prefer fly fishing. There are some days and places where it's very difficult to fish a fly and spinning is the only way to search the depths; but there are many other days and places where a fly will out-fish a spinner. To paraphrase Hugh Falkus, there is no such thing as an unsporting method, only an unsporting angler.

So, let's go back to the Findhorn up there in northern Scotland. It's March but feels like January in Reykjavik. I'm walking across a sandy bay (yes, real sand!), but the sun is weak and there are ominous clouds rolling in. I have a Spey line with a fast sink tip and a conehead Willie Gunn. I start to cast slightly down-stream and across, and get into my rhythm. There's no need to wade, as the river is not overly wide. Ice is forming in the rod rings as I retrieve line at the end of each cast. The snow has now begun to fall. It's beautiful and we are dressed for it, so not feeling it too much yet. After half an hour it's so thick that visibility is down to fifty metres. A quick adjourn to the hut for coffee (no dram for me these days, as I've already said), and gradually the snow relents and we can resume fishing in earnest.

I'm fishing a shallowish run and occasionally feeling a rock or two as my fly swings round. The next rock feels different, and I slowly lift the rod. Never strike a salmon on the fly! A miracle has occurred: I am attached to what is, for me, the ultimate prize, a spring salmon. This fish was extraordinary, bright silver with a hint of blue on its sleek head and a faint hint of lilac dusting its flanks.

That fish was caught using fairly conventional tactics. But there are other ways to catch spring fish. In recent decades a fly or lure has emerged called the Sunray Shadow. Its use is sometimes

regarded as controversial because you cast it out fairly square and fish it round very fast, often by pulling the line in quickly by hand – stripping, we call it. Some snobs seem to think this is akin to spinning (see Falkus's comment above). But the Sunray Shadow is nothing new. For decades spring fishers have used a pattern called the Collie Dog (not very imaginatively named – it was originally tied with the hairs from the tail of the owner's Collie). It's a long fly, four inches or more of hair on a simple aluminium tube, and fished fast and fairly high in the water; so not a lot of difference from Sunray. There are some specialist tactics for getting down to spring fish, but there has been a tendency in recent years to fish flies a bit faster, often imparting movement and stripping or tweaking the fly from the moment it is fishing, rather than just in the last couple of yards and then retrieving in the slack water before casting again.

I was lucky enough to fish the River Naver once (never asked back!) and catch my largest spring salmon in a howling sleet storm. I was in King Lear mode, 'Blow, winds, and crack your cheeks', singing into the sleet as I cast and worked the fly from the off when, WHACK! No need to slowly lift the rod here. That fish weighed twenty-one pounds and was a bar of silver. I can't remember what I was singing when it took my fly, nor will I ever know if it took out of appreciation of my dulcet tones or out of fury at having to listen to such a cacophony. But I do know that I didn't notice the cold at all. What a fish!

So, rather than give advice, I'll make a suggestion for when you start out on your spring fishing. Whenever possible, fish with confidence, especially if there are signs that fish are in the beat. Imagine that each cast is going to produce that incredible slow

draw. Pick your favourite fly and fish down fairly convention-
ally, deep and slow. If you want to, try another cast from the
same spot and work the fly differently, but don't hang about long.
When you get to the end of the pool put on another fly with a
different profile, size, density or radically different colour. Go
with what local advice recommends and fish down the same
stretch again. If you've got time, go back again and fish down
with a big fly fished fast. Bon chance!

TROUT AND THE MAYFLY

John: I've never had much luck in the mayfly season, which is
why Paul can tell you much more about fly patterns and strat-
egies. A lot of my young life was spent in the North West, where
mayflies are not exactly at their most exuberant, or in Norfolk,
where the tiny chalk streams can see them prolific but erratic and
increasingly scarce as the decades have passed. And when I have
ventured down to the Wessex rivers in the right season, the
mayfly have been present but not spectacular. I have always
missed out on their glorious best, always been tantalised by tales
of their glorious past. I have only the one mayfly moment which
is worth the telling, but at least it underscores the impact these
glorious creatures can have.

I have to go back to my university days for this episode, on the
River Glaven, a ten-mile-long chalk stream that enters the sea
at Cley on the North Norfolk coast. A mile up from the sluice
gates that separate salt from fresh water is the fourteenth-
century bridge at Wiveton village, and above that the trout

fishing was, at the time, day ticket and well managed. For reasons of cost and excitement, my attention was generally on that 'lost' mile beneath the bridge, water that rose and fell with the tide, where the fishing then was free to anyone local and where you never knew what the next catch would be. Big roach, pike and perch jostled cheek by jowl with brown trout and, in high summer, colossal sea trout. Add the occasional carp, tench, sea bass and flounder, and what a place it was for a lad with both bait and fly in his mind.

In all my years there I never saw another angler apart from village boys slightly younger than me around the sluice gates. The banks were wild, lonely, unmanicured, and difficult to access over marsh and across bridges so small they could have been built for fairies. The water was never deep, unless on a spring tide, and almost always clear, which is why, in September 1970, I saw a truly big brown resplendent on its lie halfway between the bridge and the sea. The more I peered, the more convinced I

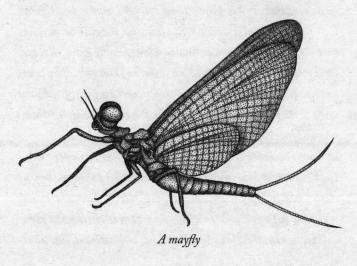

A mayfly

became that this was a resident brown and, as such, a much more wily foe than a sea trout skipping in and out with the tide. I tried it over and again with one fly pattern after another until the end of that particular season, but never did the fish as much as turn its head.

In the Easter holidays, 1971, I was back home, and my first day was a pilgrimage to the river. This sounds like a tale, but, yes, my brownie WAS there. Same lie. Same obduracy. Nothing I put over it even caused it to shift or bat an eyelid. Each evening I got myself there, as those cold dusks sneaked in, figuring the low light would help me. Bollocks to that, said the trout. The days passed, university called, but this time, late May, I was back home for study week. The Reformation and the English Church's break with Rome? I knew that all I'd be studying would be that trout.

You know what's coming already, and it's all too simple. I creep down one warm late-spring evening. There's a dribble of mayfly coming off, and the target trout makes it first mistake and in its greed sips in my Green Drake fly. A Titanic struggle and there you are, a giant in the net at last. But it didn't happen like that – not quite. There was an abundant mayfly hatch that late spring evening and, yes, I did put a Green Drake over 'my' trout on several occasions, but with no response whatsoever. It was not until a stickleback took the fly and the trout lifted itself and smacked into the stickleback that the evening unfolded in bizarre fashion. I reckoned the hook was a big enough one to do the job, a 10, I'm sure, so I waited five seconds and pulled into the biggest trout I had played since my lost Tintwistle fish (back at the start of the OUR FLIES chapter). This time I knew what I was about and landed the fish three hundred yards downstream, just above

222

the sluice gates, the place where later that summer I was to lose my biggest ever sea trout. That, though, is another story.

What to say? I kept quiet about the details of the capture when I paraded the trout in the George later, but I would never have got near the fish if it wasn't for the circumstances of the hatch – the lesson being that mayfly simply drive rivers crazy. The fact that this particular fish was nigh on the last I ever killed for either food or pride is another, secondary lesson. The fish weighed just shy of eight pounds, and it looked a sad thing dead on the bar compared with its vibrant, living self of a couple of hours earlier. I barely ever visited those marshes again; the river there had lost its allure for me. A story of success? I think not, in the end.

Paul: *As John says, the spring mayfly phenomenon can be so monumental that it becomes the high point of an angler's season – and, yes, I've had some extraordinary and successful sessions with them. But too often I get caught up in 'the heat of the hatch' – those times when the air is a snowstorm of these magnificent insects and the trout are hurling themselves skywards in an orgy of gluttony. You see huge fish here, there and everywhere, and all your focus and discipline goes haywire. To be honest, I've had better, more controlled sport when there have been hatches of hawthorns and sedges (not as dramatic, exactly, but less bonkers).*

Yes, the peak of the mayfly hatch can just be too much; just a surfeit of action. I'm just as consumed by the mayfly as a creature in its own right, not just as a catalyst for OTT trout fishing. The mayfly is the insect that thrives everywhere a river is pure, that has this fascinating lifecycle and that everyone notices when its explosive spawning cycle takes place. The years that the mayfly

spend hidden in the silt as nymphs is such a contrast to the spring
and early summer, when the adult insect hatches and rises into
the air. There's such beauty and poignancy in their story. The fact
that they don't feed and have no stomach. The sadness that their
lives can be counted in hours, a couple of days at most. The brutal-
ity of that short existence when they are prey to fish, hawks,
herons, snacking otters and even toads. What a glorious insect –
its wings of gauze, its three proud tails, its mating dance that
heralds its death. It's a story that Shakespeare couldn't write.

I have a belief that mayflies provide the origin for all fairy
stories – the greatest hatches occur as the sun sinks and often you
watch great gyrating columns of mayfly rising and falling in the

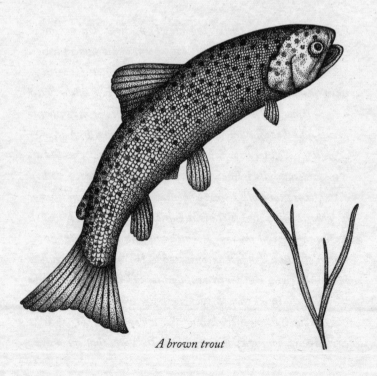

A brown trout

golden light. There's a haze, a halo around them, for all the world painting them as some diaphanous being from the underworld. A good while back I took Bob to the river during the mayfly season so he could see the wonders of nature writ large. He'd largely fished rivers trotting till then, but this day his fly casting improved to the point that I left him to it. He put on a mayfly imitation and, all by himself, cast out, rose, hooked and landed a cracking trout. He did it all, and that has fixed the magic status of the mayfly forever in my mind.

CHAPTER TWENTY-ONE

SUMMER SESSIONS

This is so approximate that a statistician would be grinding teeth, but here goes. Very roughly, if you break down the percentage of fish we catch season by season over the course of many years, judging by diaries, photographs and simple reminisces, it breaks down a little like this:

Winter: 10 per cent
Spring: 15 per cent
Summer: 45 per cent
Autumn: 30 per cent

This isn't an exact science! We fish less in winter – Paul certainly so. And, of course, autumn floods and winter freeze-ups cause havoc to rivers and stills. Spring fishing comprises largely of salmon fishing, but trips are short and these days a catch is far from likely. Early autumn can be golden, but November is often tricky, with heavy rains and early frosts, so fish numbers might hold up in September and October but often plummet thereafter. But summer, ah, there we all go. Halcyon days of warmth and plenty that see the fish feeding hard. Yes, summer is the great season, but that does not mean there are not pitfalls and periods of heartache when the fishing is far from easy.

Droughts. Those periods that are increasingly common when rain becomes a memory and rivers shrink to nothingness. We remember a summer on the Exe at Cove when temperatures soared over thirty degrees and the river dawdled within bleached banks, its flow dwindled to a trickle. How do you describe the clarity of the water then? 'Gin'? 'Crystal'? 'Pellucid'? It was like the trout were hanging in air, and those that weren't were hidden under rocks and crevices. What to do? What can you ever do in these devastating periods? But forget fishing, let us worry about our fish. Oxygen can fall to worrying levels, especially at night in the shallows of still waters where the plants absorb whatever is available. This is when the stressed, sickly or aged fish are most likely to find life an unequal struggle. It's at this point that the rivers' gravel shallows are at their lowest and most at the mercy of other water users: canoeists and wild swimmers can create carnage by destroying spawning redds and uprooting essential weed growth; the weather grows hotter and hotter; no rain falls; the dew burns off long before breakfast and, although thunder crackles in the dead of night, no storm breaks the furnace-like conditions. The land is parched, pastures turn brown and crops wither, saved only by irrigation that dewaters the rivers even further. This is a Saharan nightmare.

Cold, damp summers have their impact, too. Waterfowl suffer, as wind and cold conditions mean wet fledglings cannot get warm or dry and whole broods fail. Low pressure badly affects feeding fish and fry can starve to death in water that has less natural food than a decently warm summer would provide. Fish are vulnerable after spawning, and this is when their struggle becomes a crisis. Winds from the north and east suppress all life, and old fish find neither sufficient natural food nor the inclination to feed when they do.

But enough of all this gloom! Mild, overcast days with intermittent showers can see sensational feeding spells, especially in a brisk westerly breeze. Gentle rain that ceases in the small hours can stimulate wild activity around dawn. Steady rain in the late afternoon can drive fish food crazy, especially when that rain drifts away towards sunset, leaving a close, clammy night. There might be epic fly hatches, especially of adult common gnats emerging from their pupal cases, and what a stimulation this proves to be. The violent rain of a thunderstorm can be your saviour, especially if the weather before has been baking. The downpour will reduce water temperatures to a more comfortable level, increase oxygen content and even give clear water a tinge of colour while washing worms and terrestrial insects into lakes and rivers.

So, bring on summer! You might have to get up early and stay out late. You might even sleep the short night through by the water, but make sure you watch out for the midges and mosquitoes! Summer is the time for spotting fish as they laze in the sun-drenched surface layers of any water. Stalking is at its best and floating baits are taken with abandon. Even at its worst, summer is the best of seasons: most fish caught, most fish seen and most methods cajoling you on. No wonder the school summer holidays were the best times of our lives: barely sleeping more than five hours a night; eating nothing but sandwiches since early July; and turning the colour of conkers by September. We walked, cycled, bussed, trained and fished, fished, fished till even we were fished out and longed for the football season to start!

OUR GREAT BARBEL ADVENTURE

To help run a river – well, a part of it – is an amazing experience in our lives, something neither of us would have dared dream of when we were kids just fishing wherever we could find to cast a line. As we are now riparian owners (*Paul: 'How fancy is that?'*) we have a measure of control over how the river is managed, but that privilege comes with great responsibility, something we are slowly understanding and growing into, we hope. There is plenty of public access and that is fine, providing everyone takes on board that a river, even one as resplendent as the Wye, is a living thing, fragile, vulnerable and dependent on loving care, like the rest of us. Whether you canoe, swim, fish or anything else – responsibility is the key message of the future for every one of us who uses or lives by a river.

The Wye! The epitome of a river that is as God intended – not dredged, not straightened, not constricted by weirs and mills, not hemmed in by factories and housing developments. A summer dawn here and the sounds through the mist come from the chuckling water, the waking birds in the forest opposite, and the sploosh of fish moving heavily in the half-light. It's exciting, unsettling – we feel like a leopard could materialise from the shadows, brush past us and be swallowed up in the first rays of the sun. (*Paul: Like my nights way back on the Welsh River Cothi after sea trout, when the locals told me about the panther loose in the valley. Undeterred I might have been, but there was a panther behind every bush in my mind's eye and even the owl became some terrible beast.*)

The big fish that break the pewter surface of the river this century are barbel – hardly ever salmon, the species that made the river's name and fortune back in our grandads' time. The salmon run clings on, but so feeble is its grip that we feel almost loathe to put up a rod for them. Not so the barbel. These are strong, robust and, in the Wye, seemingly indomitable. To hold one is to feel what rock-hard muscle is all about. Rigid as a poker, with its fins erect and colours glowing, a barbel is a fine quarry – one we yearned to tangle with back in our days of worshipping *Mr Crabtree*. (*Paul: Don't think those days are quite done with, eh, John? I guess* Crabtree *will be with us till the last!*) There's a drawing in *Crabtree* that crystallised the barbel dream for us both. It's in black and white – an angry, arrogant, bull-shouldered barbel, hooked and 'going off with a heavy rush and boring deep'. We both came across it in our respective childhoods and we never believed that we'd ever see such a fish. And that was how we felt as the ink dried on the river's title deeds.

We knew barbel had to be there. We couldn't be so unlucky as to buy the one stretch of river without them. But they still took some finding. Like our imagined friend, the leopard of the morning, they appeared to us only with caution, little by little as the summer ticked by. We've said that dawns taught us a lot – especially those still, muggy ones when barbel love to roll like 'golden brown cats', as one poet wrote. As barbel are shoal fish, a barbel breaking surface traitorously gives his brethren away to any angler who's about early and takes the time to see. That's how we found one barbel lie, but there had to be more. We just knew it.

The idea that's long been held is that an angler can tell a barbel swim through watercraft alone. But we're not sure that's entirely true. Barbel choose to live anywhere, in slow or fast water, in deep or shallow water, over beds of gravel or silt. They simply have a mind of their own, and if we think back to a hundred favourite barbel swims there are just as many differences between them as there are similarities. On our beat, we realised that we would just have to find fish and use our eyes to do so. For this we needed the sun to crest the wooded hillside to the north of our bank and stream into the river so that the water would open its secrets to us. Once the light was beaming down over our right shoulders, then we knew that with polarised glasses, and sometimes binoculars, we would find more bars of gold, more barbel treasure. Waiting for sun is never a problem if you have a hut and the means to make a cup of tea. To sit on a warm bank, the river glistening, with a mug and a mate is as good as finding barbel. (*Paul: Almost.*)

First, we found barbel on the Wall, the remains of an old salmon croy, one of those jetties the salmon anglers built a century ago so they could have better access to well-known lies. The Wall, like so many croys before, has largely collapsed, the victim of a lifetime of floods and frost, but the boulders and bricks that made the structure remain, splayed out along the bed. Here the barbel hide, and one steaming July afternoon we stumbled upon them in all their balletic glory. It just takes one flash, one gleam of gold as a fish turns and its flank reflects the sun in a dull gleam to tell you that here is another shoal. It's an ecstatic moment, a hug-each-other moment. Seeing is believing and, like a veil had been lifted, before our widening eyes were some seventy or eighty barbel. The fish were visible to the naked eye as they turned, but through the binoculars we could pick out individuals, their fins, their scales, their very eyes. What a moment of complete exuberance. Deep down we both knew that this was a prelude to catching fish, but we doubted whether that would trump the actual finding of them. We've continually made the point that fishing is as much about natural history as catching fish, a belief that's central to us both.

It takes three days of feeding to convince uncaught barbel to take bait. In the Wye, the fish naturally eat invertebrates, bloodworm and small fish, largely minnows and gudgeon, which are present in endless numbers. Boilies, hardly traditional but effective, are alien to the river, but first the chub power on to them and the barbel sit back and watch, until they too come to the feast. It might take time and five kilos of bait at least, and expect the first bites to be tentative, but once the barbel start investigating then the fireworks won't be long coming. Though again, we have to say that just watching those pulsating fish coming to the loose-fed bait was quite enough for us and made the thought of hooking seem somehow irrelevant.

By mid-August we had spent a lot of time on the water, located six large groupings of barbel and weaned them onto our baits, which they were taking with gusto. We knew we would fish, but we had become proprietorial – or would paternal be a better word? You cannot change the way fish think and behave without taking a measure of responsibility. By mutual consent, we agreed on rules. No, principles: we would never take more than one, at the most two barbel from any one group in any one session. A single hook-up would satisfy us and, above all, such restraint would not send the shoal into meltdown. We'd refine our methods, too, we decided. We'd stay completely mobile. Rod, reel, hooks, baits, net, chest waders, and off we'd go, each time starting at the head of the beat and working our way down to the bottom, six hours' fishing if we really wanted to stick at it. (*Paul: Something we never actually did.*)

Wading (safely, of course) should never be seen as the prerogative of the game angler. Getting into the summer river is an immersive experience. You see the water from a different angle and in different dimensions. The gravel beneath your feet and the current of water against your thighs takes on real substance. You are in with the minnows and, amazingly, as you fish, so the fish will come to you. Repeatedly, as the summer peaked, we drew barbel to within a rod's length of us, sometimes to our wader boots. This was spectacular for us, a sport within a sport.

When you are wading, you don't need the apparatus that bogs so much angling down. You don't need a box, bait tins (a pouch around the waist will do) or even a rod rest. You hold the rod for as long as you are out there, and this method is called touch ledgering. What a liberating way to catch big fish. Cast, let the bait settle, point the rod tip to where you think it is lying and hold the line between reel and butt ring in the fingers of your non-rod hand. We could tell you about the pick-ups, the way the line 'buzzes', vibrates, twitches, pulls tight and often snaps into your flesh hard. But the point is that you will know; your instincts will assume control. This is a skill that relies on the heart, not the head. The rush of blood when the first barbel is hooked is a corporeal tsunami. Within a day, the experience is addictive. All you want to do in life is feel that connection between you and the fish. The line becomes an umbilical cord, a link to the very pulse of the barbel's behaviour. A barbel can deliberate five minutes or more before making its decision. It commits. You feel it. You strike. The rod hoops. The bottom explodes. That becomes enough; those few seconds you want to recreate again and again.

We urge you to take this from our book above all else. The splash factor in a low river is a killer. Throw out a heavy feeder or lead and the thwack on the surface can empty a swim of barbel and chub, too. One cast can finish off your chances for an entire day. Occasionally the weight of a 15 ml boilie was enough to hold bottom in moderate flow, but generally one, two or, at the most, three SSGs spaced at foot-long intervals up from the hook did the job we wanted. And those SSGs entered the water almost exactly like a scattering of boilies, a sound to attract, not repel. But Part Three: The Nuts and Bolts of What We Do will tell you more.

It's impossible to describe the freedom that fishing feather-light can bring, when the dividing line between game and coarse fishing becomes blurred. And on a few occasions it was fly tackle that we took with us. John's wife Enoka had a barbel a shade under ten pounds on an artificial nymph fished on a light fly rod, and our mates James Buckley and Pingers had fish just a little smaller. It could be that the day we do without bait altogether might be approaching.

We have said that barbel choose where they live, but we'd happily agree that a steady current, a clean bed and shelter in one form or another are all important ingredients in the making of happy barbel. And if there is one thing they like above all it is rocks.

We've discussed the old croy, but there is another barbel hotspot where we have access and that, too, owes its existence to ancient history. This time we are harking back to the era of the Romans, when the Wye was an important artery as they tried to push ever west into Wales. Here the river narrows so that further boat access proved impossible; two thousand years ago this is where the Romans built a port. Fragments of fossilised timber remain, but the stone that the Romans used is still there on the river bed in vast quantities. These are blocks too heavy to be rolled away by centuries of Wye floods, and now they harbour not boats but a huge shoal of barbel. John found the place and its scores of fish back in the early nineties, but it is more recently that the historic background has been researched and understood.

There is, however, a snag: the river here is so boulder-strewn that fishing is nigh-on impossible. There is one small pocket towards the head of the long run where the rocks are small enough to allow careful bait placement, but elsewhere too many snags abound. Even if the tackle does not catch up, then a hooked fish always does, and we can't live with that thought. The lacking of fishing matters not a jot to us. This is a place where it's good to sit when the sun is high and we can watch the barbel roll and flash with gay abandon, rejoicing in the delights of the warm river.

SUMMER CARPING

John: Paul has encouraged me to remember an old friend, Rob Shanks, who tragically passed away a few years back. The enormous number of people at his funeral told of his generosity and his warmth as a family man and friend, but it is as a carp angler that his story should be mentioned now. The carp anglers who Paul and I revered in our young days, men like Richard Walker, Jack Hilton, George Sharman and Fred J. Taylor, were famous names then but all but forgotten about now, and it would be good if Rob's legacy could last a little longer and we could learn from it in these pages.

I've already mentioned that Rob had a young family, but he also ran a thriving tackle shop with his partner Dan, which allowed him some time on the bank, fishing for his beloved carp. Long-stay sessions were difficult for Rob, so he relied on short visits packed with imagination, strategies, concentration and cunning plans. He could often accomplish in four hours what it took less-focused anglers four days to achieve, and it was my privilege to often watch him in action. Never was there a finer demonstration of his art than a summer evening at Charity Lakes in Norfolk. Charity is a pretty place, with its three lakes and stretch of river all set in woodland. It is not in the least bit exclusive, but rather the sort of venue you or I could access and enjoy for a pretty reasonable sum. This makes for tricky fishing, particularly for carp, which know anglers' wiles inside out. To catch a good fish from Charity in a few hours in full daylight would never be considered an easy mission, but it is one I saw Rob pull off several times.

I was on my way to the river, and Rob invited me to spend a few hours watching, talking and perhaps doing a little fishing. He had his gear set up but he expressed some doubt that he'd be using it. Everything depended on what he might see. We walked together around the main lake, slowly, carefully, deliberately, and Rob marked fish that I absolutely did not. I fancy myself as a fish spotter, but I was cut down to size then, for sure. A single bubble. A solitary twitch of a willow branch where it met the surface. The slightest vortex of moving water. Just a tiny stain of brown, indicating that silt had been disturbed. Rob noted all of these signs, but one clean patch of gravel the size of a dinner plate particularly caught his eye. It was all but hidden in a tree fringe, but I saw it plainly when Rob pointed it out, three or four yards out in five feet of water. Rob broke a dozen boilies into flakes and threw them so that they settled on the gravel. We could just make out the dark pieces of bait against the paler background. And then we waited.

At 8 p.m. the sun was slanting across the lake, and quite clearly a carp of good size appeared, paused, tipped and went down on the gravel for a minute or less. We stopped our soft conversation and Rob was a figure rigid with attention. The carp established a pattern of drifting in, going down on the bait and sauntering away again. For me this was excruciatingly tense, but Rob remained a pillar of cool. He flicked out a boilie and let out slack line so it was lying perfectly in the taking zone. It was now a little before 9, and when the carp came back a fourth and final time, the line twitched and tightened enough for him to strike and for mayhem to break loose.

We didn't weigh the fish but both of us put it at twenty-two pounds, and with a high five Rob was gone. It was job done and time for a 'good night' to his kids and an hour with his wife. If dear Rob were alive to read this, he'd dismiss it as nothing unusual. He'd point out that there are many carp anglers with busy lives who still catch good fish. He'd add that many long-stay carpers have a similar affinity with the fish and can read the water in magical ways.

Much of the 'professional' carp scene is not for Paul and me anymore, but that does not mean that we do not recognise some of the superb fishing skills out there. One glory of fishing is that there are so many entirely different ways to enjoy it and so many different species and waters to explore.

In July 2021 we had as good a carp experience as it gets – this time together. Picture a two-acre lake on a beautiful afternoon, when the carp were everywhere on the surface, ploughing the lilies and basking the day away under the willow fronds. Even for us, after all our years in the game, it was a mesmerising scene; if you had told us as kids that we'd *ever* fish a place like this, then 'died and gone to heaven' would be an understatement. Rod, reel, size 4 hook on the line, unsliced loaf and landing net, and, along with glasses, this was all the kit we needed. We marvelled at the fish here and we drooled over them there. All were crackers, generally fully scaled commons, and the loose samples of bread crust we threw out to them were found, inspected and nearly always greedily taken with much sucking and slobbering. We were mesmerised!

What we wanted to find was that special fish, a real belter of a carp, and in a place where we felt confident of landing it. There's little point hooking a whopper in a lily bed where you stand zero chance and all you do is stress the fish … and yourself. Then we found it. A wondrous common had a short patrol route from thick cover around the island into the middle of the lake. It would hang around there for a few minutes, taking in the sunshine before heaving back into shelter. We reckoned a piece of crust landed close to the island, smack in its usual route, stood a good chance of being taken.

Paul cast out the fifteen yards. Perfect. He straightened out the line and then sank it by dipping the rod tip under the surface. This would avoid any drift and help keep the crust nailed where we wanted it. We waited. A kingfisher zipped past like a halcyon dart. We waited. Small silver fish pecked at the crust, making our hopes leap. We waited until 'our' fish showed itself, dark and menacing, an inch under the lake's surface. Slowly, slowly it edged into the crust zone, and we waited. Closer, closer. 'Bloody hell,' whispered Whitey. Lips emerged. A whole round mouth. The crust was engulfed with a slurp that drowned out the roar of our heartbeats.

The carp sank, the crust had disappeared, but Paul held hard and, crucially, waited for the line to draw tight. Pandemonium. The peaceful afternoon was whipped into a watery frenzy. How that carp fought, but Paul did all the right things and kept that carp in open water. Try as that fish might, Paul would simply not let it make sanctuary by the island, and his gear held tight. Within ten more minutes the fish had been landed, quickly photographed and released. Unforgettable!

John: There's not much more Paul and I can say about carp fishing in the 2020s. The whole scene has changed beyond recognition since our young days, and we don't want to come across as Luddites lamenting a lost past. Back then we had scout tents; now there are luxurious bivvies. Back then bite indicators were cylinders of silver foil; now electric alarms are high-tech as space capsules. Back then bait was bread and potatoes; now boilies sell in their billions. Back then carp waters were secret, few and far between; now commercial carp waters proliferate all over the UK and Europe. Back then a prized carp weighed ten pounds; now a 'thirty' is a moderate target. Back then an angler served years of apprenticeship with smaller species before graduating to carp; now kids start at the top with all the gear, the bait and fish like whales before them. Back then carp anglers were a select few; now the carp tackle industry caters for millions of them worldwide. None of this is bad. It is certainly irreversible, and we accept what has happened with an open heart.

We've also made the point that there are some very fine carp anglers fishing today, and Alan Blair is a great example. I've done a good deal of work with Alan over the past few years, and while there is a vast gulf in our ages, spiritually I like to think we could be twins. I've watched Alan's burning enthusiasm for the sport. I have admired his technical mastery. But most of all it is his imagination and his ambition that excite me. Alan sees fishing from so many angles and exploits so many of its possibilities. All his sessions are a whirlwind of action and initiative. He faces every challenge with bravery and positivity, and nothing in his fishing is too off the wall not to be tried and tested. There is nothing fixed in his thinking, nothing stereotyped in his approaches.

Perhaps what I identify with Alan most is his passion for all species. The Special Sessions Alan and I have shared have sometimes been chasing carp, but just as often the target has been tench, bream or crucian carp. Only recently, we fished a shudderingly cold day for barbel. We knew that, given the temperature, we stood little chance, but Alan still fished with ferocious concentration and only gave up when the line froze in the rod rings.

We've talked perhaps too much about our past heroes. Izaak Walton and Hugh Falkus have featured repeatedly, but it's good that there are modern heroes that we regard with equal respect. We do feel that fishing today has become more predictable, but it is good to know that there are young anglers determined to make it as thrilling as it ever was.

THE SUMMER SEASHORE

John: We've said from the start that we'll only write what we know about, and sea fishing has never been the main focus of our angling lives. That's especially strange, since I lived on the Norfolk coast off and on for twenty-five years, and Paul and I have both done a fair bit of saltwater fishing overseas – for bonefish in particular.

In the very early days I had a stint digging lugworms for a living and then holding a few back for the beach at night with my mates. There were still good cod to be had, like the night I was out on the storm-battered shingle and me, along with my dearest friends of those heady days, Billy and Joe, landed an eighteen-pounder between us. It glowed like butter in the lamp light and it seemed a shame to kill it, but food for free was always a tenet of the countryside back then. 'We'll cut it into thirds,' Billy proclaimed, and that was the fate of the biggest shore-caught cod either Paul or I have ever lived to see.

We also had a passion for grey mullet around this time. We fished hundreds of hours for these shadowy creatures that come and go with the tides, following the skinniest water, feeding on God knows what. The only time we caught any at all was in nets, and when the fish were cleaned all that could be found in their stomachs was algal-weed-stained sludge. Given that these North Norfolk marshes covered thousands of mud-covered acres, how could a globule on the hook ever be found and taken, we both reasoned? Convinced of the futility of the quest, we gave up for thirty years.

The sea-fishing challenge was abandoned till I found my life on the coast again, just after the millennium. From the attic window of my cottage, I could watch the sea and the lagoons behind the sea wall through binoculars and mark the gull activity that told me the bass, mackerel and mullet were back for the summer. On an evening I took to walking to the beach with a light spinning rod, catching a mackerel or bass for a supper as fresh as you'd ever find. The mullet evaded me entirely until one dawn in September 2006.

I arrived in the car park at Cley around 4.15 a.m., took out a fly outfit, cameras and binoculars, pulled on my chest waders and set off to walk west towards Blakeney Point Reserve. The sky was lightening behind me as my boots crunched along the gravel spit, and the sea was as serene as a mirror, with barely a cat's paw of breeze. Already it was warm, and long before I reached the Watch House, I removed the hat that the cool of dawn had prompted me to wear. By 5.30, sunrise was approaching, the sky a wash of lemon and yellow, and the harriers were circling above. The silence was immense, the sea barely kissing the gravel spit and the tide noiseless as it spread over the marshland lagoons to the south. Seals were busy offshore, hunting bass and mackerel, and the creeks and muds of the marsh were filling with the tide. Slowly at first, the mullet came into this skinny water, until they swarmed in their hundreds and the bay that I had identified the previous day was churning with bodies. I put up the rod and tied on a shrimp pattern finished off with a red tag on a size 14, just as the magazine experts advised. But I had no faith. How many wonder-mullet claims had I read or heard about over thirty years or more? This was sure to be another fantasy dreamed up to sell a thousand hopeful words. Still, the appearance of the sun and the chorus of tits and warblers gladdened my heart, and who cared about a mullet on a morning like this?

The water was up to my knees as I stepped off the heather into the fast-filling lagoon. Mullet were all around, feeding hard, forcing my heartbeat just as they had back in the days with my cod-fishing comrades Billy and Joe, way back in the last century. Even as I cast, I knew it was no use, that the fly, the latest mullet miracle, would be ignored like all the baits and lures of the past. After a hundred casts and retrieves, fast and slow, the light was blinding, the tide was on the turn and the mullet were slowly slipping back out seawards with the receding water.

I was thinking about home when, with a zip like lightning, the fly line was tight, the reel was scorching and a mullet of seven pounds was cartwheeling in the autumn air. On and on the fish fled till the backing was out, and with a final flourish and a jump of three feet or more ... it threw the hook. The whole affair had lasted forty seconds, but it was a love affair, a meeting ecstatic in its violence and passion. Who cares about a hook hold? The simple fact that it had been taken was the triumph of the morning. A Special Session in so many ways.

Paul: *Don't know why John is making all this fuss over mullet – 'fool's sea trout', I've heard them called. Forget all the angst, John. I've caught loads of them on a handmade rod on the Greek island of Eos. I bought some line and found some bamboo canes growing in a garden. I asked if I could harvest one, whipped the line on the end and had endless early mullet mornings on bread baits. The seas were beautifully clear, and you'd see the mullet*

terrorised by great bass or wolf fish that hammered into them
and then picked up the morsels of the battered prey. It was pure
drama, with me an audience of one.

Bass, happily, are fish with a mouth wide open and eyes for a
meaty meal. Both of us have had our bass moments, some of
them on *Gone Fishing* duty. Remember when Paul and Bob
went out in the West Country in a storm that blew them out
of their hats and was enough to sink ships? The skipper gave
them no chance of anything until Bob hooked a blinder of a
bass that pulled and pulled and had them all forgetting to be
seasick.

It's fine to have a bass from a boat, but the best and simplest
way is to fish from the shore. Wading in shallow, clear water
with the bass weaving their magic around you is the way to
do it. There are endless hundreds of miles of bass beaches
around the UK: Cornwall, Wales, the south and east coasts,
Ireland and even as far north as the Hebrides in these warmer
days. Just keep safe. Ask advice on the tides and avoid adverse
weather. Go your first few times with an expert. Keep warm
and dry. A life jacket might be the best investment you make.

Spin with silver spoons and plastic fish patterns. Try
topwater plugs. Fly fish with streamers and even pike flies
that are all fluorescent and flash. Watch for feeding gulls and
gullies and tide rips where the current is at its quickest. A
boat can land you on a sand bar where you can walk, exploring the drop-offs. Try rocky headlands, sea pools, beds of
weed and kelp, old pilings, jetties and any structure that
provides a feature in the vastness of the sea – piers can be

brilliant early in the morning, before the tourist day begins. Work around sea defences and groynes, and keep investigating anywhere and everything that could suggest food, protection or both to a bass. The sea is big, but it is not faceless. Watercraft, that oft-used word, might be more elusive here, but the sea still gives out clues if you are open to reading them. When you get it right and your rod kicks to its first bass, when that old fishing adrenalin surges, we both know how good it is to be by the sea.

Paul: This shoreline stuff is exactly how I dream about bass fishing. I've never done much, certainly not as much as John, who lived by the sea for years, but it's always been a mission of mine, something locked in my heart and imagination.

When I have these bass moments in my head, it will always be a silvery dawn, with the sun rising and the sea a mix of pearl and pewter. The wind of the day will still be hushed, and bass might even be striking at smelts or sprats, making the sea boil. I can feel the adrenalin, the pulse of excitement as I push out a fly line and retrieve a big, bushy Muddler or something silver back towards me. That hammer blow. That magnificent creature all aglow with shiny white scales and bristling fins. The sea's flawless creation, if you like.

Trouble is, I know I'm never going to do it, or at least not in my foreseeable future. Is that a trouble? Perhaps these fantasies are what fire us up, counsel us through the tough times when we're stuck in the doldrums of life. The other day, John told me about an amazing perch session he'd enjoyed with our mutual friend Simon. I love perch and these were crackers. They'd had getting on for twenty big 'uns in the day and, even better, they had watched the striped monsters chasing fleeing shoals of bleak right by their floats. It had been like taking part in an Attenborough film. I'd just love to be a part of that experience, and even though I almost certainly never will, it's like I can live it vicariously and get my own slice of joy out of it. That was the drama of my old fishing adventure books when I was little. I never thought I'd have the money or the knowledge to catch the fish my heroes did, but I was still there with them in my mind. When I put down the book and switched off the light, I had my dreams ready and raring to go.

SUMMER TREATS

John: Paul loves the big rivers – the long, probing cast, the moment when, thirty yards off, the surface shifts and the line pulls tight to a serious, silver salmon that's so special it takes your breath away. There's no denying it, that's the cream of the sport. But the small rivers in the summer have a magic that enchants, a suspense that tingles. Arguably, there's no better destination than the West Country, that peninsula of heath and sinuous valleys where rivers like the Exe, the Barle and the Dart weave their way. We know that runs of salmon are not what they were, but there's nothing new in that sad statement and this isn't quite the place for anger and lament.

Paul: Though awareness of how our salmon have suffered these past forty years never leaves our consciousness.

John: Yes, very true. Watching you fish the Exe at Cove showed the fragility of the game, and though you fished like a champ, you stood no chance given the stifling heat and water low and clear as a small gin before the tonic's added.

Paul: Incidentally, my biggest salmon came from a dead low river in conditions like these. It was hot, bright and the sun was burning on the low water. I knew it was hopeless, but you'd never fish if you always waited for the perfect time. I tied on a small yellow two-hooked fly and fished through the glare with no hope when a clonker took it. I never weigh fish but this had a length-to-weight ratio suggesting it went thirty pounds. Bugger me! It just goes to show, you never know.

Paul's always good to watch, entering the river like a heron that has a world of experience hidden in those hunched shoulders. Wading with barely a ripple. Casting long and delicate, line a web in the light. Total concentration, gaze like a gimlet where a grilse has shown itself early morning. That's the thing about Paul, about all top anglers. They'll fish hard, even when deep down they know they stand little or no chance. You just never know – lightning has been known to strike from as clear and blue a sky as the one above us now. What we both know is that this is early June, and the last days of August could be far better … *if* the rains come. What we also know is that it needs to teem down on the high moors and the rain gush into the valleys where it will fill the rivers with warm, brown rising water. Then the salmon will run from the sea in their swarms, eager to get to their spawning redds way upriver. On their way, these spirited fish can create angling memories to last forever.

What you need is the luck to be on a river like the Barle at exactly the right moment, just when the high water has peaked and is beginning to fall ever so slightly. A single-handed rod will do, a 6 weight, we'd suggest, with line to match and a tippet of perhaps eight pounds breaking strain. Small flies are what you want, perhaps tied on double hooks, size 8, 10 or even 12. Then it's a case of walking and exploring every last nook and cranny where a salmon could be paused and taking a breather on its journey. The deep pools are obvious, as are the runs into the heads of these pools. But there is nowhere you can rule out. That eddy behind a rock. The deeper runnel under the alder branches. A push of water funnelled between two boulders. The slack behind a fallen tree. Even a cattle drink, normally shallow but now offering a deep, calm slack.

To get the best from the day you should start at first light and not finish till dusk. That's a twelve-hour session and you'll need a pint, a pie and a bed at the end of it, but you'll sleep like a baby. You might walk ten miles or more and fish fifty places for five minutes each if you do it right. This is hard work but the best fishing is just that, and if you hook a fish on a West Country river then you'll know about it. In fast water especially, there's no need to delay the strike; there's just a pull and a fast-retreating fish. And if the salmon has got down those rapids, then you'll never bring it back up again. Breathable waders, a wading stick and a brave heart all help in this summer spree, but what a privilege it is to be in pursuit of fish like these in the twenty-first century. Of course, they'll go back in the river after unhooking and they'll survive fine because they are small and fit, and the distances they have to travel are not immense.

Remaining on summer rivers, let's look at a target a little more accessible – our old mate the chub. (*Paul: My best chub is six pounds-odd from the Lea. It fell for a sprat meant for a pike, but it just shows what a fisher-friendly species they are. They'll always have a go. I won't hear a word against them.*)

Is there a river south of Hadrian's Wall that doesn't hold a chub or two? Is there another species that is so obliging in the stifling heat? Is there any other fish that can be caught in so many ways? A dry fly; a nymph; a small lure 'popped' across the surface, throwing up a wake as it goes. Natural insects, too, along with worms, slugs, crickets, expired frogs you might find and dead mice. (Never clean out a trap and throw the mouse away! They float, and chub smash into them drifted down on the surface. We accept that might not be for every-one.) Also, maggots, cheese, corn, boilies, pellets, meat, sausages and almost anything you might find in the fridge that can go on a hook. Lures of every type, both sinking and floating; baits float-fished and ledgered. But best of all, our all-time favourite and what catches us most summer chub in the most exciting way: freelined bread, either slow-sinking flake or surface-presented bread crust. This is how we do it on rivers from the Wye to the Wensum, from the Trent to the Test, and everywhere in between.

Get yourself a bucket and two white unsliced loaves. Soak thoroughly and then mash everything up into a sloppy goo, which is largely mulched bread but with bits of crust left intact. Find a nice steady length of river – on the Wye a glide a hundred yards long is perfect. Ladle out a couple of handfuls of the mix and watch it go down with the current. If it's warm with a gentle breeze and the river is clear, you might see looming shapes emerge subsurface and gobfuls of the bread disappearing fast. It could be that lower down this glorious glide, the river is exploding as more chub attack the floating crust remnants with a piranha-like voraciousness that's hard to believe.

A twelve-foot rod will do it, or a quivertip rod – or whatever you have got, to be honest. A four-pound line is about right, and on the end tie a size 6 hook. That's all you need for this most immediate of angling action adventures. Do you start with floating crust on the hook or sinking flake? Perhaps go up top and see how things develop, but you've got to realise that crust is harder to present naturally and those wily old chub are easily spooked by any tiny thing they suspect is not right.

A piece of crust the size of a conker, then, the hook pushed into the white underbelly and through the hard brown stuff on top. Dunk it to give a bit of casting weight and to soften it so the hook pulls through on the strike – if there is one. Cast it out, and now the hard part. You have to make sure the crust drifts down the river exactly like the free offerings you have thrown in. If the line gets pushed about by the current and forms a big bow, then you'll get drag on the crust and it will get pulled off course. That just will not work for you. The chub will follow, nose the crust but will never take it unless you straighten the line and iron out that bow. This is the gentle but critical art of mending, and it's why we advise a long rod that makes the whole process easier.

If you see chub stationed fifty yards beneath you, creep down halfway so you only have to control the crust for a shorter distance, but take care. Disturb those chub, which is easy to do, and they'll simply melt off downriver. Get it right, get a take, and you have to hold your nerve. The chub will sink with the crust in its lips and think a little while. If it likes the taste of it it'll move off, taking the crust in as it goes. So, when you see your line slide across and tighten, that's when you strike, hard but steadily, with no jerks or erratic pulls. And keep that strike going, right over your shoulder if necessary, till you feel your fish. You've got a lot of line to pick up, remember, and a big hook to sink home.

We reckon that you do well to catch more than a single chub on crust because these are ultra-sensitive fish, wired for survival, and the commotion will have the whole group on red alert. But their downfall is that they are greedy, too, and going the flake route might very well winkle out a second fish for you. Take a big fistful of soft flake from the white insides of the loaf, a piece the size of a mandarin orange is about right. Push the hook through till it pokes out the other side. Hold the hook by the bend and squeeze the bread around the shank so it's nice and secure. Then drop it in the margins and make sure it sinks ever so slowly … perfect. Lob it out into the flow of the river, where it will splash like a small sheep falling in. Let line out so the bait drifts naturally down the water column and along with the current, and chances are the line will tighten again. Remember what you did with the crust. Wait till the line is direct to the rod tip, then strike with controlled power! Fish on. Two chub, both big ones, and it's almost always job done. Now almost certainly you'll have to find new chub, as that group of fish will have scattered. You might get to locate half a dozen chub shoals, catch even eight or nine fish, walk three miles and have the time of your angling life.

OUR MEDALLIONS OF GOLD

There was a time that we both remember (*Paul: So that's a long time ago!*) when every farm had a pond, every meadow had a marl pit and all of them held small but perfectly formed crucian carp. It was in puddles like this that all country children learned the angling art, all by themselves, making their own mistakes, finding their own solutions and, best of all, making their own magic. Summer life was all cow parsley, cuckoos and crucians. That was then, and by the new millennium all the crucian carp waters had vanished and the children along with them.

Crucians are our original carp species, present in Britain from the Iron Age and probably very much earlier. They are small, three pounds is a monster, but less is more and they are perfectly formed medallions of gold.

Much of the crucian problem lies with the agribusiness that has taken over from traditional farming. Tractors don't need water like horses did, so the farm ponds dried up.

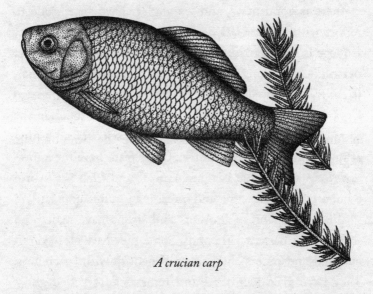

A crucian carp

Modern fertilisers are used uniformly, so the marl pits fell into disuse and neglect. The agricultural workforce shrank until there was no one to clean out ditches and field ponds, which gradually choked and died. The hamlets grew to villages and the villages to towns, and crucian waters were filled in, levelled out and increasingly built upon. More roads, more chemicals, more run-off and more diffuse water pollution followed these developments, along with greatly increased otter predation. We have to report anglers' part in all this, too. Larger mirror carp were stocked in historic crucian strongholds, and the smaller species found themselves bullied to the edge and beyond. Just as painful has been the hybridisation that takes place when discarded goldfish are released without thought for the indigenous fish present in a pit or pond.

There is good news, and we anglers have an episode to report with pride. Carl Sayer is a professor at University College London, a Norfolk farmer's son and an ardent fisher person. Carl, along with students, academic colleagues, friendly farmers and anglers like the exemplary Bernard Cooper, have turned the whole crucian crisis around, in the east, at least. Here we see the marriage of science and fishing at its happiest and most fruitful. Carl's team have the know-how to establish what is a pure crucian and what is a hybrid – important in validating and spreading genuine populations.

This brave band of brothers and sisters have revitalised scores of former crucian ponds. Some they have cleaned out by hand and others they have opened up with machinery. They have even identified and reinstated so-called ghost ponds, hollows that fill with mist on a summer's dawn and indicate the location of a crucian home long ago. Once there were pools waiting to house the fish, Carl netted crucians from waters where they were not at risk and transported them to fill the vacuum. With astonishing speed, these new arrivals settled in, reproduced and provided juveniles to reseed more waters that were coming on tap. A quarter of a century ago, it is no exaggeration to say that crucian carp were on a red list; today, their recovery has been nothing short of triumphant. There's a massive lesson here, we both think. Carl and his crucian crusaders have used top fishery science to magnificent effect. A problem was identified and a solution was quickly formulated. Expertise and hard work joined forces, and with virtually no money the crucian project powered forward. No pointless data collecting or years of endless meetings and debate, just volunteers with a plan and the will to get on with it full-bore.

> *John: Crucians are cunning critters, and even when they are small they can be an angling nightmare. Take my experience on a small pool about the size of a tennis court at Lawn Farm. I was tasked to evaluate the crucian status there and fished well for three evenings in succession. They were balmy, mellow dusks and I watched my float go under repeatedly once the sun had dipped. It was all roach, though, and not a crucian showed itself, not on the hook, not on the surface, 'not nohow', as one of the tractor drivers put it before cycling homewards. When Carl's team netted the following week, the fyke nets (those long tunnel contraptions originally used for eels) were hauled in the following day bursting with crucians. It was a biblical catch, big enough to make Captain Birdseye blanch. Not a single fish weighed more than ten ounces, but they were there in their thousands – more fish than water, you could say.*

As you might expect, the bigger and wiser crucians grow, the harder they get to fool, if that is possible. On waters that see any serious angling pressure, we'd say they can become next to impossible, certainly if you are wanting to do it right and get yourself a crucian on a float. Dawn is absolutely your best bet right till the end of the summer, but do be aware that that period before sun up can have its hazards. Even at the height of a warm spell, 4 a.m. can be perishing, notably if a wind of any strength is coming in from the east – when you might as well stay in bed, if you want our advice. If it is one of those close, muggy mornings when the light just seeps in then grand, *if* there aren't midges or, worse, mosquitoes clouding the air. If your crucians are living anywhere near marshland, do take the necessary precautions of a hat, a head net and plentiful DEET!

These days, to say that there is a 'right' way and a 'wrong' way to catch a fish is to risk being called 'purist', 'old fashioned', 'fuddy duddy' or far worse, but we don't care. The fact is that the best way to catch crucian carp is on a float, and on many waters so-called improved, modern methods have made this almost an impossibility. In fact, the only way John knows how to crack a crucian of size on a float is from a swim shallow clear enough for him to actually see the bait being taken. Picture this: under the lilies, the water is shaded and cool. A sliver of sweetcorn lies on the bed four feet down, just visible from above. Two dark shapes loom into view, very hesitant but moving calmly towards the bait. One of the fish tips up slightly, its tail rising in the water column. It hangs for five seconds above the corn and, in the blink of an eye, the bait is gone and the strike is instantaneously executed. Crucian on. Job done. Home to bed and breakfast. A glorious morning is breaking and a triumph of summer fishing has been achieved.

CHAPTER TWENTY-TWO

AUTUMN SESSIONS

We love Indian summers – those beautiful, mellow days when the warmth seems all the more precious because we know it is on the wane. Some afternoons it can seem like true summer has never left, but there are hints that life is changing around the waterside. Early mornings are now damp and misty, and spiderwebs hang like hammocks, weighed down by drops of pearly moisture. Your boots still get dew-drenched at midday, and a blue haze hangs over waters deeper and thicker as the year progresses. Days are shorter, sunsets blaze and are gone in the blink of an eye, while the chill drops down from a clear sky before your gear is packed away. The trees still cling to green but the tall willow herbs have burst their seed pods, smothering the banks with a clinging down. The ditches and dykes are carpeted with withered water parsnip and loose-strife, and the burr marigold is nigh on the only flower in bloom.

It's true that through all these changes water temperatures grow less stable, but there are still damsel flies perching on float tips from mid-morning and the shallows remain rich in mosquito larvae and water fleas, so all fish species are feeding well. Early October is just about the best time of the year to mount a bream quest (*Paul: Sorry again, Bob!*), because in part it is dark early and a session lasting between 5 and 10 p.m. is likely to coincide with shoals going out on their patrols. These feeding spells can be hectic if there is good cloud cover, a mild wind and a full or new moon. These two lunar phases are caused by the sun and moon being in alignment, and their gravitational pull causes inland waters to rise and fall in a seemingly infinitesimal way, but enough to impart huge energy and appetite to all our fish, bream and carp above all.

Predators are also making the absolute most of every feeding opportunity open to them. Fish that are intent on food are easy prey for pike, but the season suits perch above all. The shoals of smaller fish that perch feed on need to work harder for their own food as the banks of water fleas decline sharply, and this forces them to travel further, often in open water, where they become more vulnerable to attack. The sheltering weed beds are dying back too, and even the subdued light of autumn disguises the perch as they prepare an attack from the encroaching shadows.

Make the most of these golden days, we say, because soon come the frosts in their full fury. Within days, the waterside world changes abruptly. The warblers desert the reed beds, the wagtails are leaving the ditches, and the swallows, swifts and martins are long gone, along with the whitethroats and wheatears. Apart from the oaks, trees are shedding leaves in earnest, and these rot and sour the water with their dust and traces of chemical toxins. The lilies fall back, the water clears out, the whole world of fish becomes more empty, more cold, more hostile. Temperatures descend further, and some mornings the margins are coated with cat ice till the sun is well up. This is when fishing takes a huge knock, the more so when blue skies, bright suns and high pressure are dominant. Forget fishing in the day if you can and get out at dusk for an hour or two, when there is some glimmer of hope in the darkness.

Now come the gales of late October and early November, and they arrive with such ferocity that every remnant of summer is ripped out and pulled down by morning. The trees are stripped, the rushes are flattened and the lily beds are torn apart and tossed on the windward shore. Accompanying rain is generally heavy, cooling still waters even further and raising rivers to flood levels. We are now staring true winter in the face and the migrant birds have arrived. We anglers might now expect to see wigeon, pochard, geese and, of course, those dreaded flights of cormorants coming in from the east.

Our message? Fish hard September and October whenever you can, because they see the best sessions of the whole year. Don't write off November but remember that every kind day is a blessing, and sooner or later autumn will give way to winter. Looking back at our diaries we see that over the past four seasons our last barbel of the year have fallen between 11 and 19 of the month. That's hardly any sort of rule, of course, but then fishing isn't ever an exact science and how fervently we hope it never becomes one.

AUTUMN PERCH

John: 16 November 2022 and I witnessed a phenomenon on the Wye that took my breath away. You reach a time in your life when you think there are no surprises left for you, but you are wrong. Nature never exhausts her power to amaze.

The river was summer low and very clear, and air temperatures were high for the time of year. I had spent the day with dear Frank Hall. Frank is an old-school angler who always puts the welfare of fish first so that, after a barbel or two, he was content to let them react and we decided to fish the late afternoon for perch in a deep hole off the main river. By 4 p.m. Frank had taken three fish of moderate size when, right by his float, chub of half a pound or so sheared out of the water, pursued by very visible, very large perch.

So far, so interesting, but what followed was extraordinary. For forty-five minutes similar explosions erupted everywhere up and down the river. We had a view of at least a quarter of a mile of water, and this large beat of river was constantly pockmarked by ferociously hunting perch packs. In fact, 'constantly' doesn't remotely do justice to the frequency of these attacks, as they happened every five seconds, here, there and everywhere our amazed eyes could see. I climbed a high bank where I could look down on a favoured feeding area, and at one point counted seven groups of perch, each half a dozen fish strong, all pursuing chub of up to a pound in weight. This was a tiny fraction of the water being affected, so there must have been hundreds, if not thousands of perch taking part in this annihilation of the chub. And how *big were these fish? I know you want to ask. Most must have been three pounds, but many looked to be more like four – and more. Then, just before 5 p.m., the frenzy stopped as abruptly as it had begun and the river oozed its quiet way south.*

Frank and I failed to capitalise spectacularly, but to be fair we were caught wholly on the hop. Ensconced in the Old Salutation pub afterwards, we thought we should have fished six-/seven-inch lures, cast far out and brought back fast just subsurface. Big popping plugs and chuggers could have worked, or perhaps large deadbaits retrieved sink and draw against the current. Our worms sitting in the silt hardly made any impact, and I hope to be better prepared should I ever witness the like again.

What was going on that autumn afternoon? Throughout this book Paul and I have tried to link our fishing approaches to the natural scheme of things. We always aim to think fish first, and bait and strategy second. There are those who fish the same bait, fly or lure wherever they go, irrespective of conditions, but we feel that's putting the cart before the horse. I told Paul exactly what happened on 16 November 2022, and we suggest that on the cusp of cold weather a huge number of smaller fish were beginning an exodus downriver towards Hereford, where they traditionally spend the winter. Why they do this might have something to do with the warmth of urban environments, but the issue is that these tens of thousands of fish were up in clear water and uniquely vulnerable. Could it have been that huge numbers of big perch were actually accompanying them on the journey, striking into them on the way? During the downstream migration do more and more predators join the procession, until silver fish, perch and pike alike find themselves down in the city

*reaches? Of course, it is conundrums like this that keep the
angling passions stoked and make every single session special and
unexpected.*

Autumn perch are creatures to behold, the colours of the
season seemingly reflected in every shining scale. Big perch
waters can be divided into natural ones and commercials, or
artificial ones. The first category would include the Scottish
lochs or waters in the Lake District. We are also going to add
the very long-established reservoirs like Chew and Grafham
Water and perch-famous rivers like the Thames, the Ouse
and the Wye. The big reservoirs have held big perch since
their construction, and that is down to the number of prey
fish they hold; because the reservoirs are so large, angling
pressure rarely becomes an issue. We have made it a rule not
to talk about approaches we know nothing about, and much
reservoir perch fishing is from boats using lures at great
depth. This is not where we are comfortable, though we have
had big perch from reservoirs like Upper Tamar in the South
West. Here we fished from the bank using deadbaits in a
swim we baited for roach, knowing the perch would be sure to
follow. Paul landed a fish as close to three pounds as dammit,
and there were half-chances missed during the session.

When it comes to larger waters, we have both fished
Kingfisher Lake in Norfolk when perch there were in their
prime. As soon as the cold weather set in, the majority of the
small roach on which the perch fed forced themselves into
three bays on the north-eastern side of the pit. Here they
spent every winter and became the target for huge numbers

of pike and, of course, large perch. It was common to watch both predator species striking into the silver fish shoals throughout the day, and the perch could be caught on float-fished worm and deadbaits as well as lures of all sorts.

However, it was quarter of a mile from these bays, where a bridge leads to the island, that we really had our fun. The stanchions of the structure were a great draw for the perch, and here, too, the inflow from a drainage dyke entered the lake, which was rich with very small roach, chub and dace. In crystal water six-feet deep, it was possible to watch big perch coming and going once the sun was on the water, and we caught them on a waggler-float fished overdepth and two lobworms on a size 6 hook. We rarely used any shot on the line, preferring the weight of the lobs to fall through the column and provide an anchor in the underwater currents. Heavy baiting with maggots around the float concentrated the silver fish and, of course, drew in the perch. The clarity afforded us amazing sights, especially the speed at which a three-pound perch can consume an eight-ounce roach. A simple flare of the gills and the prey is hoovered into that vast cavern of a mouth.

Our fellow angler Tim called out that he had a very large perch hooked from his position on the bridge, and we ran to help. In the crystal water we watched as a splendid fish fought deep with powerful, plunging runs. Tim knows his fishing and all was going to plan when a colossal perch emerged like a fantasy from the shadows of the bridge and engulfed the perch on the line. Spellbound, we saw the leviathan turn the smaller fish and swallow the head, but it struggled with the fiercely erected dorsal fin. Inch by painful inch, Tim tried to guide both fish to the waiting net, but with only a yard to go, the larger fish shook its head and the smaller perch swam free, only to be landed and weighed at a fraction under three and a half pounds. We all agreed the lost giant had to have been *at least* twice the size of the fish it attacked. There had been stories by the carp anglers of a perch in the area that we would not believe, and we had been stupid enough to discount these tales. What a lesson painfully learned, and though we fished there many times over the years, the leviathan never again showed itself.

River perch have undoubtedly grown larger this century as the American signal crayfish has spread exponentially. John first witnessed two very decent perch attacking, dismembering and devouring a large signal way back in the nineties on the River Kennet, and this phenomenon helps explain more recent captures of four- and five-pound perch from the rivers Ouse, Wensum and many more this century. It's easier to locate big perch on rivers than on big still waters, and perch holes tend to be inhabited season after season. This fact does help in knowing where to fish, but the more you know about perch, the more you realise just how edgy and sensitive they become over time. You wouldn't believe those great, bony mouths of theirs can be so discriminating, and that those big eyes of theirs can be so capable of spotting flaws in our presentation. If a big perch feels any resistance as it takes a bait, it will eject it in a millisecond, way before any strike stands the slightest chance of connecting. Big perch fishing in the autumn is a highlight of the year. Treat them with respect, fish for them carefully and you'll catch memories to last you all winter long.

Neither of us fish small commercials a great deal, but we are very aware of their perch potential and the fact that they have produced record fish in the past. In many of these waters, perch live under the radar and you have to sniff them out, because most of the anglers here are after silver fish or smaller carp. That's an advantage in many ways, too. Perch that are left alone lead stress-free lives and are infinitely more catchable, but we also know that if you do find perch in a commercial of a few acres, there won't be many of them. Once you start catching the same perch again, we'd suggest you move off and look for waters anew. Big perch have telltale scars of a tough life all over them. A tattered fin. An old cormorant scar. Blind in one eye. Every fish has its identifiable wound, and once you come to recognise these, it's time to go. Enjoy your time. Catch the half-dozen fish resident, then move on, carrying your perch secrets with you, is what we suggest on these pocket-handkerchief waters.

AUTUMNAL TREASURE

As October teeters its way towards November every session becomes special, tinged with a fear that it might be the last of the halcyon outings before winter sets in. Every gust of chill wind spells menace. With every frost, leaves are lost and the all-important water temperatures fall another degree, and we know that we are fishing on borrowed time. Every cast, every strike, every fish becomes even more special, and all our fishing takes place in a landscape that could have been painted by Constable himself. Autumn and the whole gamut of angling possibilities are laid out before us. Golden afternoons are perfectly designed for rudd, if you can still find them in today's UK. We used to know lakes rich in them, like gleaming plates, they were, but, sad to say, otters have done for the big ones and cormorants have mopped up the smaller ones.

More likely are some last-gasp tench, again late-afternoon, when the water is at its warmest and mirror still. Scatter corn or maggots in the margins, sit back and wait for dark shapes gliding in to investigate. Set a waggler float overdepth so that six inches or so of hook length is lying on the lake bed; watch for tiny bubbles to rise around it and strike when you see it dip away or shimmy more than an inch or two right or left. You'll only get one, such is the disturbance a tench makes when hooked in shallow water, but move along the bank and you might get a second chance before the light goes.

Carp and barbel are both feeding hard as the cold weather approaches, and sometimes the biggest fish of the season comes as October gives way to November. Sunny days are great for both species, but if the night that follows is clear and there is a trace of frost, you might have to wait till lunch at least for fish to feed positively. If you can, arrive as the sun is sinking and fish through dusk into the first couple of hours of darkness. This is electric stuff, but wrap up warm, as temperatures can sink like a stone. You'll notice carp and barbel both deepen in colour come late in the season, and scales that were golden have become copper and then mahogany. Just another reason to cherish the autumn.

Bass will still be off the beaches, certainly through October, and that's when chub will be voraciously active, hungry enough to chase topwater flies or lures as well as taking every bait you can throw at them. Once again, dusk is the top time, but when the mornings are pewter grey with mist, you can think pike for sure, as their season is beginning all over again. They'll be very active way into November, but there'll still be prey fish around so perhaps give them a big lure that really commands attention.

Then, out of nowhere, the rain will turn to sleet, the frosts will settle hard and, before we all know it, winter is here again. Another angler's year has completed its circle. If you listen to us at all, you'll realise that this is the beauty of being an all-round angler. We can understand the person who *only* fishes for carp or salmon or pike or whatever, but we think they miss out on a lot by doing so. It's their choice, of course it is, but we've learned to love them all.

YOUR OWN SPECIAL SESSIONS

Of course, you'll be making your own sessions, successes and memories, and we'll sign off with a few final suggestions. Make sure you experiment with different methods, approaches, skills and venues. Diversity is a common clarion call these days, and it's very relevant in fishing. *Gone Fishing* has tried to showcase the amazing variety of fishing that there is out there and point out the glories of a varied fishing life. You'll see Bob as happy with a fly, float or Flying C salmon lure, and that's what makes a happy angler, in our view. We also recall September 2021 on the Eden, filming the *Gone Fishing* Christmas special. We found a heart-stopping run of a hundred yards, all chuckling water between waist- and chest-high. There simply never was a more succulent salmon stretch, and Paul positively purred. He fished it with confidence miles high, and that translated into an exquisite session with a silver fish to show for it. He was in his absolute element – and an 'absolute element' is something we can all find, if we look for it.

There's one point here, though. In the same way that we have suggested creating yourself a signature method, we would agree that there is a whole lot of benefit in knowing a home water exceptionally well, a place that you really know how to read in all its moods. Recently, John visited a dear old friend on the upper Wye. Philip has fished the beat here for years and knows every single rock, pebble and salmon-holding lie. He can gauge the exact conditions and instinctively predict where running salmon might hold up and from where they might be tempted to take a fly. Years of experience have coalesced into the profoundest knowledge of these three miles of water, and his results are arguably better than anyone else's on the river ... though with typical modesty he'd deny that adamantly.

And *yet* ... there is also an argument for fishing as many venues as you possibly can with as many anglers as you can meet. We have talked very little about angling overseas, but we would both like to emphasise that we have learned life-changing lessons from fisher folk in other countries and even other continents. We've experienced some amazingly special sessions, but also brought home approaches that we've used to great effect in the UK. Perhaps, if time and opportunity allow, knowing a few waters intimately and visiting many more waters whenever possible are the perfect recipe for a happy and productive angling life.

We have tried to build fishing into the fabric of our lives in innumerable ways. We've mentioned how hectic full lives can seem, but we have learned that even the shortest sessions can be special if they are intense. A Sunday morning before the family is awake can be rewarding, especially between June and September. Those Smuggler rods have a purpose. Travel rods like these break down into eight sections a foot or so long and can be packed both for business visits and family holidays. We've fished in New York's Central Park before the streets are cleaned, and as for Venice ... travel rod in hand, you'll never see a sight more wonderful than the sun rising over the Gulf of Venice. Take photographs. Write diaries. Never let the uniqueness of the sessions be forgotten and you'll lead an angling life of contentment.

PART THREE

THE NUTS AND BOLTS OF WHAT WE DO

We have told a lot of stories and revealed a great deal about how we fish and what fishing means to us, but what about the nuts and bolts of the sport? Part Three, then, is a handbook of the basics and methods that we actually employ when we're out there doing it. You might well be gobsmacked at how simply we fish. We hope so. We have little truck with a lot of the modern gobbledegook we read or hear about. We've been learning everything that follows since the 1950s (*Paul: Sixties in my case, Guv!*) and it has all worked for us so far. We write with open hearts and hope this might help you on your way.

In the interest of clarity, we've arranged this information alphabetically, and we'd like to think that any potentially confusing technicalities in the text will become clear now (*Paul: Some hope*).

Beats – simply a slightly posh name for a stretch of river that you are fishing. Commonly applied to game rivers, but not exclusively, in our view.

Binoculars – these are terrific aids to watching fish and allowing you to see what's going on out there. You'll see what flies are hatching, and if you are surface fishing for carp you'll be able to see those lips come up and the biscuits disappear. Buy the best you can afford and don't be fazed by all the size options – 8×32 or 8×40 are perfect for the angler. The other tool you'll need is a good pair of polarised glasses (see under 'P').

Blockend feeders – the type of feeders we commonly use fishing maggots for tench and bream in still waters. Take off the lid at the top, pour in the maggots and replace the lid so that it is enclosed. The maggots dribble out of the holes in the sides. We fish blockends fixed on the line generally between six and twelve inches from the hook.

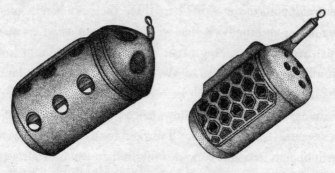

Blockend feeders

Boilies – we hate them but can't ignore them. Since the 1980s the top bait for carp, but now used for tench, bream, barbel, chub and even roach in smaller sizes. A bit like marbles, they come in multiple colours, flavours and sizes. We like brown and red boilies for tench and barbel, mostly 12 or 15 ml, and chub will take pretty much all of them.

Bolt rigs – again, an invention of the carp world now used throughout bait fishing. The idea is a bait, commonly a boilie, is fished on a hair (see 'H') and a heavy fixed weight is placed on the line. The fish picks up the bait, feels the resistance of the lead, bolts in panic and hooks itself as it flees. Effective but not purist.

Bread – a proper old-school bait that's hard to beat. Flake is best. Simply pull a piece of fresh white bread out of a loaf, push a size 10 hook though it and mould the bread around the shank, leaving a nice fluffy portion around the bend. Crust is great for surface fishing for carp and chub. Bread paste is excellent, especially mixed with ripe cheese for chub.

Bubbling – tench and carp disturb the bottom when they feed, often releasing gases there that break on the surface in the form of big bubbles. Smaller fizzy bubbles are also created by air passed through the gills of feeding fish. In either case, a real giveaway and one of the top signs that watching water can offer you.

Carp and carp fishing – when we were young it was hard to find a carp water. Now it's impossible to find a water that doesn't hold them, often to the detriment of other species. Carp fishing is the big thing for many; it's where the money in the sport lies. Much of it centres around boilies, bolt rigs and long stays in bivvies. John did a lot of it as a kid and even wrote a book on it, so we're not being sniffy. Some of the best carp anglers are genius. Most big carp are either fully scaled (commons) or have scattered scales (mirrors.) Most clubs have carp lakes and there are many day-ticket waters. If you are really keen you might graduate to join a group of anglers, known as a syndicate, which rents or buys a water for exclusive use.

Casting – if you are considering fly casting then we urge you to take proper lessons from a qualified instructor. John taught himself to cast and even now has ingrained bad habits that he would not have adopted had he been taught properly. Casting bait or lure is largely a matter of practice. Look at where you want the bait to land. Don't let the bait, float, feeder or lure get too close to the rod tip, and we'd advise always having the weight you want to cast out at least two to three feet beneath it. You'll achieve better accuracy and fewer tangles as a result.

Catapults – the tool we used as kids, now useful in getting baits out long distances with accuracy.

Centre-pin reels – most of us use fixed-spool reels that make casting any distance easier, but centre pins are perfect for trotting a float on rivers and for close-in fishing on still waters. Many would dispute the advantages, but if you look at the image you'll see that you can place your thumb on the spool of the pin and that gives you great control when playing

A centrepin reel

a big fish. Lift your thumb if you need to give line, but clamp down if you want to stop the fish running. A pin takes time to master, but with a well-matched rod we truly believe you can't beat them in the heat of battle.

Clothing – we are not here to treat you like plonkers, but if you are wet, cold or broiling hot, you'll be uncomfortable, unhappy, and you'll go home. There's so much great wet, cold and heat-wave gear on the market now that there's no excuse for being underdressed on the bank. You'll need waterproof coats, hats and trousers. You'll need good waterproof footwear. You'll need thermal underlayers for winter. You'll need good hats, something with a brim for the sun and something warm for the cold. We'd advise looking into chest waders. We'd advise having a complete set of dry cloths stored away in your vehicle (if you have one). We'd advise buying from recognised outdoor suppliers rather than from the tackle trade. Sorry, but you get better gear for your money, unless you go to one of the top game brands like Orvis or Simms, which are both great in quality and design.

Cormorants – these fish-eating birds have vastly increased in numbers this century, as many have begun to flock in from Eastern Europe from November onwards. Many rivers and lakes have lost huge numbers of smaller fish as a result, and it's blindingly obvious that if you don't have small fish you don't have big ones. And for the future of fisheries, it means fish that don't grow to maturity, don't spawn and so numbers spiral downward. There is a certain fallacy going around that our greedy needs have stripped the seas of the fish that cormorants require and that, as a result, the 'poor' birds have been driven to hunt on inland waters. Rubbish. This century the number of freshwater-feeding cormorants has rocketed, and these are birds that used to live in Eastern Europe but now spend the winter here. Our native cormorants are largely marine feeders, and their numbers have stayed much the same during this period.

Corn – sweetcorn was discovered as a fantastic bait in the seventies and remains brilliant for carp, tench and bream in still waters, and barbel, chub and roach in rivers. Use a grain on a size 12.

Czech nymphing – as the name suggests, this is a brilliant way to present either a single nymph or a team of two or three nymphs to trout or grayling … and it originates from Eastern Europe, too. One key to the method is that you use a small strike indicator on the leader. This is a tiny float, in reality, and you set it so that you can fish your nymphs at any level in the water column. It also registers takes immediately, too, hence 'strike' indicator. This is a super-efficient method in rivers, primarily if you are wading. You use a very short line and flick the flies upstream of you. Let the current bring them down in front of and then beneath you. Raise your rod, flick upstream and begin the whole process again. Search the ten yards or so of river like this and then move position.

Dapping – a highly effective method on big, still waters, generally from a boat. It is especially useful in the mayfly season, and the heart of the dapping scene is on the big loughs of Ireland. You drift with the wind, letting a real or an articulate (artificial) mayfly pattern bounce on the waves before you. A long rod is used, often six yards in length, and the line can be silk or floss, something that catches the breeze and helps the fly dance in front of you. Very exciting! Especially when a trout takes the fly with a gentle slurp. Do *not* hurry the strike, but instead lift firmly into your fish after a second or two. This can be a slightly stomach-churning method in high waves. Don't be afraid of suggesting it's time for a Guinness or two.

Deadbaits – all predators, pike especially, pick up dead fish from the bottom, so they are scavengers as well as hunters. Pike take dead freshwater fish and sea fish with equal excitement. Our top baits are dead trout, roach, sardines, smelt and small mackerel, but most fish types are worth investigating. It can pay to flavour the baits with fish oils or even dye them, though we don't bother and we catch plenty of pike.

Eddies – an eddy is a phrase from the watercraft lexicon, and it is simply one of those areas off the main flow of the river where the current is comparatively still. Eddies are especially 'hot' in times of flood, when all river species like to get out of the full force of the water. You'll often find them behind fallen trees or downstream of bridges. Fish them with a float or look for bites on the rod tip. Try luncheon meat for chub and barbel, lobworms for perch and bread for roach. The **Crease** is where the slow water of the eddy meets the quick water of the main current. You'll see it quite clearly, and fish like to sit there dodging in and out of the flow to intercept food and then rest a little while. Or at least that's what we *think* they're up to.

Electronic devices – over the last thirty years electronic fishing gadgets have increased with the speed of a running tarpon. Now you can buy endless fish-finding devices that give you depths, position of fish and the size – perhaps soon even its name. Bait boats can deliver boilies for carp or dead fish for pike at any distance you care to fish and at any proximity to snags you deem to be safe. Now there are drones to fly so you can spy on the exact location of carp in a lake or barbel in a river. Look, we'll be honest, as we have been throughout. John used sounders for ferox as early as 1986 and much good did it do him. He also used bait boats for piking more recently with more triumphant results. *But* ... we're uncomfortable with it all. Bait boats can and do cause tethered-fish problems when baits are too far into dense submerged roots and branches. Drones can be worse than irritants as far as fellow anglers are concerned, and shortcut tactics undermine old skills like watercraft, which we feel are fundamental. How far do we want to go in our desire to catch fish? Is any method acceptable? This book is called *How WE Fish*, and in large part we'll do without high-tech, but the choice is yours and the wizardry is out there.

Feeders – we've talked about feeders here and there, but they're not our favourite bits of kit. They do work, though, and get samples of bait down to the bottom close to your hook bait. We do use method feeders for tench. You pack a method feeder with groundbait, bury the hook into it so it doesn't tangle during the cast, and then fling it out, hopefully close to where you are building up a concentration of fish. Try not to scatter casts all over the shop. Make sure that the groundbait – or feeder mix, to be posh – is firm enough to withstand the cast, or all you'll do is spray everything and everywhere with sloppy small stuff. Which won't make you very popular. We

A method feeder

also use blockend feeders, which are plastic containers that you pile maggots into. The blockends hit the bed and the maggots crawl out, enticing the fish in to feed.

Flies – all fish eat natural flies at some time or another, which gave anglers the idea to tie artificial flies on hooks and use them to fool fish, trout and salmon especially. Flies are tied using fur and feather and manmade material like tinsel, and making them has become an art form enjoyed by enthusiasts worldwide. We're a bit ham-fisted so tend to buy shop-bought creations, which work just as well. Dry flies. Wet flies. Buzzers. Nymphs. Streamers. Bombers. Poppers. There are dozens of patterns, mostly quite unnecessary. Buy/tie what

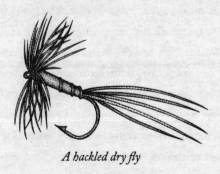

A hackled dry fly

283

you are advised that you'll need for any location and don't stress. Oh, and keep them nice and dry when you're not fishing and you'll be able to pass them on to your grandkids.

Freelining – we love doing this! It's largely a river approach, but not always. You can wade out into the edge of the flow and let a big bait trundle off downstream. Big lumps of flake are good. So is luncheon meat or a bunch of lobworms. A small dead fish. A dead mouse, even. Generally, you have no weight on the line, but sometimes an SSG shot gives better control and sinks the bait a little better. Control is what the method is all about. It's no good letting the bait waft down the river and you having no clue where it is or whether it has been snaffled. Hold the line in your fingers and feel the current pulling the line through them. A bite will be either a pluck or slow draw, or a hard pull. But you'll know. Wind down till the line is fairly tight then strike far back with controlled power. Sounds sexy, and it is. Super-exciting and highly effective. Chub, barbel and trout are all suckers for it.

Hair rigs – forty years ago, some clever carper realised that if you put a bait on a fine length of nylon or braid and dangle it just off the hook, fish will pick it up more confidently than if the bait is on the hook itself. Today, very many baits are fished on 'hairs', certainly for carp, tench, bream, chub and barbel. Boilies and pellets are the baits most associated with hairs, but you can fish corn, artificial maggots and even worms on hairs. Tying hairs is fiddly work (*Paul: 'Specially for John*), so we suggest buying them shop-tied, like most people do. Hairs are very often fished with bolt rigs (see 'B'), and are very much part of the modern scene.

Hedgehogs – we know this purports to be a fishing book but bear with us. When we were kids these fascinating little creatures were a part of nigh-on every session we fished, early or late. Then there were almost as many hedgehogs as people in the UK, and now, sitting one to a seat, there are barely enough to fill Wembley Stadium. Cars and car parks. Housing estates with squeaky neat gardens. Badger proliferations. Hedgerow destruction – the clue is in the creature's name. But, above all, we are looking at the near complete crash of insect life that has taken place in our lifetime. When Paul drove to the pub as a lad on a Saturday night, he'd spend a sunny summer Sunday scraping insects off his windscreen. Not now. That's why we've lost our sparrows, our starlings and our skylarks. That's why we'll lose our wild trout stocks, too, if we don't act fast. Anglers view the natural world holistically and see that if you lose a single piece, the whole jigsaw crashes down.

Hooks – OMG, how many types are there worldwide? There are literally hundreds, but there's no point reeling them off to make us look clever and you feel intimidated. Basically, for most fishing, you'll use eyed hooks and, if you are going very fine, spade end hooks. With the former, the line is attached through the eye and then secured with a knot (see Knots). Spade ends? Fiddly to tie, so we suggest buying them ready-tied from a shop. The big issue is the size, so a big bait means a bigger hook. A single maggot on a size 18 is fine, but for a conker-sized piece of luncheon meat you'll need a size 4.

Most of the hooks we use are between size 4 and size 18, with size 10 being great for lots of baits and fish. Buy a few hooks of most sizes and you'll be fine. **BARBS!** A big question these days now that fish welfare is rightly a big deal. We'd advise microbarbs, but some waters have rules stating you must use barbless so abide by them, obviously. For pike, trad-itionally treble hooks have been used, but increasingly double and even large single hooks have been proven as effective and more fish-friendly. Pressing down the barbs on bigger hooks makes their removal a far easier operation. Most salmon spin-ning is now either banned or only allowed using a single hook, so again check the rules. Keep all hooks dry when not in use and keep them sharp. If you think you have damaged a point, especially when fly fishing, check, sharpen or replace.

Induced take – fly fishers got very excited about this term back in the eighties, but it is simply the act of lifting a bait, fly or lure sharply in front of a fish's nose so that it is fooled into thinking its prey is about to escape. The response from most species of fish can be instant, instinctive and unthinking. Fear of losing a meal goads the fish into making a mistake. In fly fishing, lift the fly smartly at the end of a retrieve or in front of the fish if you can see it. When float fishing a river, hold the float back against the current and that will cause the bait to rise up in a tantalising fashion. Work at the technique. It's not hard and can be a real game-changer at times.

Instructors and guides – we really cannot overstate how central both these professions can be to you ... and us. If you are considering taking up fly fishing, a morning with a registered teacher can save months of heartache. In John's case, a life-time. The same is true of Spey casting for salmon with double-handed rods, but with knobs on. If you are paying thousands of pounds a day in Iceland, say, it makes sense to fish your time out as efficiently as possible. Even if you have cast a fly a bit in the past, refresher courses do nothing but good. And when it comes to guides, a good one can transform a blank day into one brimming with interest, excitement and information. If you are new to a water, a river especially, a guide can teach you so much, whether fishing bait or fly. Gear, approach, watercraft – a guide can take you to another level.

Jigging – what do we say about this? Using small plastic lures on light gear and 'jigging' them with flair and imagination has never seemed much of a skill to us. That's because we don't do it right – we don't have confidence. John should have. He saw Robbie Northman catch a thirty-pound pike on a little silver plastic jobbie two inches long and fished on feather-light braid. So, in the right hands, a very useful approach for perch, pike and zander.

Kelly Kettles – there are some things in fishing that smack of the twee: wearing a tie or red corduroy trousers, perhaps. The whole Kelly Kettle scene could be added to the list, you might think. *Wrong*. Forget flasks if you want a real bankside treat, because tea or coffee just never taste right, not a patch on the freshly brewed. There's always wood to burn on any riverbank, and if you value your friends, a cup of the steaming stuff is a way to show them. Kellys cost fifty quid or so, and operating them is a knack, not rocket science. If your other half, or you, of course, bakes cakes, that's just adding to the bliss of it all. There are two serious fishing points to be made. One, it always pays to rest water. Brewing and drinking a Kelly cuppa takes half an hour, a perfect fallow period on most swims. Two, tough days wear down the spirit and dampen the enthu-

siasm and confidence. It's far better to take a Kelly breather
and restore yourself than upping sticks altogether and going
home. (We hate to add this caveat, but given the health and
safety age we endure: don't brew up near tinder-dry grass,
crops or trees. Extinguish all embers and sparks and even
douse the site before leaving.)

Knots – it's Paul's joke and truism that John only knows one
knot, the half-blood, and has used nothing else the last sixty-
odd years. Not *quite* true, but almost. The point of mentioning
this is really, once again, to emphasise that fishing can be as
simple as you like to make it, and undue complexity isn't
always the best recipe for success. The knots illustrated in
Carys Whitehouse's brilliant diagrams are the ones we suggest
you consider mastering and which should see you through a
successful career. You should also consider purchasing *The
Pocket Guide to Fishing Knots* by Peter Owen. It shows all the

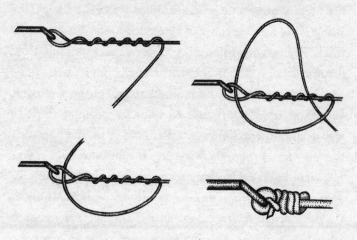

A tucked half-blood

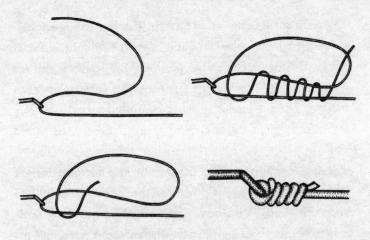

A grinner or water knot (four turns)

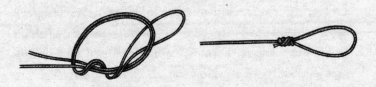

A simple overhand loop

knots an angler needs clearly and precisely. *(Paul: It's an invaluable guide. I take it with me in my fishing bag!)*

Leaders – the fancy name for the length of line that attaches the fly, lure or hook length to your main line (that is, the line that comes off the reel). Most commonly, we talk about leaders in fly fishing. The big deal is that the thin leader with the fly attached is joined to the main fly line that gives you casting weight. The leader is attached to the fly line by means of a simple loop. Take care in choosing your leader.

You want it thin enough to fool the fish but strong enough to land it when hooked. You might want a floating leader if dry fly fishing on the surface, or a sinking leader if you are using a nymph or wet fly beneath the surface. In carp fishing, a braid leader is most commonly used. We buy them ready made up, of course, whereas real experts would not.

Ledgering – simply putting the bait on the bottom, using a lead to cast it, to get it down through the current and to anchor it there so it doesn't get washed away. Leads – aka ledger weights – come in various shapes and sizes, and it's our considered opinion that most of us use ones that are unnecessarily heavy and make too much commotion. Ledgering was historically regarded as a lazy angler's approach, but obviously the rise of bolt rigs and carp fishing have now made it the most commonly used bait-fishing method. Sadly, we say. However, check out touch ledgering and see how we often do it.

Lines – obviously what we use to fish with. Fly lines are calibrated according to weight, so, if you have a 6 weight rod, a good average for most trout work, a 6 weight line marries it nicely. Still on fly, sinking lines are used to get subsurface flies down the water column, often in still waters. On most rivers, a floating line will be what you want for dry flies and nymphs fished in water that's not too deep. When it comes to fishing bait and lure, braid line is generally used for the latter and for a good deal of carp fishing. For general bait work, monofilament line is still the favourite, and you match the strength with the job in hand. We hate being obvious, but you'd use fifteen-pound breaking strain line for good carp in a weedy pool and three-pound breaking strain line for small roach in a clear-water canal.

Maggots – a bait so important that we feel it merits its own entry. You can still buy them in quaint pint or even gallon quantities, depending on how many you want. We like red for tench and white for roach in a winter river. It's a good idea to buy a mix of red and whites, but we'd personally stay clear of those dyed blue and green. Transportation? Buy a box with an escape-proof lid. Don't let them get wet or they'll climb up the box and escape; those that don't manage to escape smell so bad you throw them away anyway. Keep them out of heat or they'll die, congeal and make you sick to your wellington soles. Maggots then turn into casters before becoming flies, generally bluebottles. Casters are inert brown cases, but roach especially love them. This is the glamorous side of the sport.

Match fishing – both of us have fished matches in the past and loved the old scene in the sixties and seventies, when a bus would weave its way out of London or Stockport and take us to some exotic river we had only dreamed of. Match fishing began in the nineteenth century near the big industrial conurbations of Sheffield, Nottingham, Leeds and London, and allowed working-class anglers to enjoy their sport with mates and with the promise of a kettle or teapot as a prize. Right until the seventies, the big national matches held on the Trent, the Thames or the Fenland drains attracted hundreds, if not thousands of entries. This was probably match fishing's glorious Indian summer, and much of what happens now takes place on small commercials where putting together bags of hundreds of pounds of identically sized small carp becomes the target.

Missing bites – a bit of an odd one, this entry, but there's nothing worse than waiting all day and then missing a bite or take that you have been praying for. And just sometimes, you miss one after another, and that can lead to despair and drunkenness. Could be you are striking too fast. If dry fly fishing, count to three after the fly is engulfed. If you are salmon fishing, wait for a two-foot loop of line at the reel to draw tight. If you are surface crust fishing for carp, wait again for the line to slither out. Perhaps your strike is wrong. What you are aiming for is a steady, powerful lift into the fish. Avoid wild, jerky strikes – keep in control. Check your line is not blown all over the place by wind and that it is not pushed into big bows by the river's current. You need a direct line to make a strike count. Let's say you are fishing a big bait. You'll need a big hook to match. Sometimes varying the length of line between the hook and the lead/feeder can help. Try longer, and if that fails go shorter. Don't panic, Captain Mainwaring! Keep working on the problem and you'll get there.

Move or stick – another of the great quandaries of angling, and life, come to that. How long do you flog a piece of water that isn't producing? Do you stay because you have confidence in it, or because you have put in a fair bit of bait? Do you reckon that if fish aren't biting here then they're simply not feeding anywhere, and that you might as well stay put and wait for a turn for the better? Or can't you simply be arsed to go to all the fag of upping sticks and moving from a swim that's nice and cosy? There's no easy answer. Move and fail, you'll wish you had stayed. Move and succeed, you'll never know if the water just switched on. Stay and fail and you'll wish you'd moved. But you simply have to realise that it's always a gamble and there's no room for hindsight. Us? We're of the impatient style of angling and want to make things happen. We get bored easily, too, but above all we trust our instinct, that gut feeling that tells you what to do for the best. Note, too, that because we fish simple and don't weigh ourselves down with unnecessary gear, moving is less of a trial than if we'd brought the angling equivalent of the kitchen sink. We get this right about the same number of times that we get it wrong – but at least we know we've tried.

Nets – keep nets are still used in match fishing but few people see their need outside of that. Landing nets are a must, though. We prefer round nets to triangular ones. Telescopic handles are great – if they work. If you are fly fishing, chances are you'll want a slightly smaller net with a shorter handle, especially if you are wading. For pike, rubberised mesh is great, as the treble hooks don't get caught like they do in fine mesh. Cutting trebles out of nets when it is freezing, you have a three-pound pike thrashing about and you have bleeding fingers is no fun at all. Using the net is important. The rule is to draw the fish to the net, not chase the fish with the net. The difference might sound slight, but it's hugely significant.

Night fishing – overnight camping trips can be fun – we have already mentioned safety aspects – but don't assume noctur-nal sessions are inevitably successful. They can be long, cold and depressing. We used to do plenty of after-dark stuff, but now we have concluded that dusk and the first hour or two after dark are by far the best. Running close is dawn and an hour either side of it. You can catch between 10 p.m. and 3 a.m., but don't beat yourself up with guilt if you are safely tucked up in bed.

Particle baits – simply a highfalutin term to describe a mass of small baits like maggots, hemp and corn, rather than big baits like boilies. The idea is that if fish see enough particle baits they become preoccupied with them and turn up their noses at anything else. Works in theory and in practice *if* you put enough small baits in. Say you are chub fishing a glide in winter. It might take a pint of maggots thrown in over the course of an hour to turn them on to the feed. Spectacular when it works.

Polarised glasses – essential for seeing through the glare of the water and allowing you to spot fish. That's when you can start strategising – see 'Sightfishing/Strategising' – and plotting how to get a bait or fly to your quarry. Equally essential to protect your eyes from flying hooks, notably when fly fishing. DANGER NEVER GOES ON HOLIDAY. Buy the best you can afford (*John: Well, not always. A few weeks ago my £300 jobs revealed none of the pike Richard, my boat companion for the day, was seeing with twenty-quid knock-offs. We exchanged and, Bingo! There ARE exceptions to all our rules.*) and get advice from an expert salesperson. Ask if the frames are strong and lenses are scratch-proof. Different coloured lenses are specially suited to different light values. A good, general, all-round colour is light bronze. Tie them to a cord or lanyard so they are around your neck when not in use and less likely to be lost – or to fall in.

Pole fishing – today, pole fishing is generally a highly specialised branch of the match-fishing scene and neither of us would know how to wield a fifteen-metre pole costing five grand. But we love the history of pole fishing, most especially on the Thames, where wizard anglers held sway right through Victorian days and beyond. The great maker then was the Sowerbutts company, and examples of these glorious creations are highly sought after today.

Quivertip rods – these are rods with a very thin glass or carbon tip spliced into the top section of the rod. The idea is that these tips offer little resistance to a taking fish and show up bites really well. Quivertip rods are generally used with feeders and ledgering approaches.

Reels – we've discussed centre-pin reels, now on to the more commonly used fixed-spool reels. Carp pros use big pit reels, but for general coarse fishing we'd recommend smaller sizes, commonly described as 2000, 3000 and 4000 models. Fly anglers use reels that are in essence centre pins. Most fish are lost because the clutch of a fixed-spool reel is set too tight and line cannot be given to a running fish. *If* you are able to buy decent quality, do so. Cheap reels mean cheap clutches, and they can seize up in cold, heat, wet or simply any time … and there goes another lost fish. There are multiplier reels as well, but these are used for extremely heavy lines and for winching up fish the size of the *Titanic*, so we'll give them a miss here.

Retrieve – simply the act of winding the reel to bring the line back in. If you are retrieving a lure when you are predator fishing, put imagination and gusto into the job. Wind in a dull, metronomic way and the chances are that you won't fool any fish into thinking your piece of metal or rubber is alive. If you are trotting a river – see 'T' – retrieve your float in along the margins rather than down the line you are fishing. Always do everything you can to minimise disturbance.

The retrieve when you are fly fishing could command a chapter all of its own, but our mission is to keep things simple. If you are fishing a still water for trout using a team of buzzers (midge pupae), you should use a very slow retrieve or simply let the flies drift about in the breeze while keeping in touch with a very slow figure-of-eight retrieve. This only means coiling the fly line in your non-casting hand, and it's nowhere near as tricky as it sounds. If you are using a lure-type fly for trout, or perch or pike, come to that, you can vary the retrieve. Maybe some long, steady pulls followed by a few short, sharp strips. An old acronym (initialism?) for this was devised by an expert donkey's years ago: FTA. This translates sweetly and elegantly as Fool Them About! The main thing to remember is to experiment and use a retrieve that suits the fly. A buzzer retrieved at thirty miles an hour won't be taken, but a Sunray Shadow fished at Mach 2 will be hammered by a salmon.

Rigs – the name given to the business end of tackle, often carp or pike tackle. A rig might include the leader, the lead, the hook and the hair. You might buy rig bins or holders, boxes wherein you can store your precious rigs safely away from the tackles that dominate the likes of John's tackle bags.

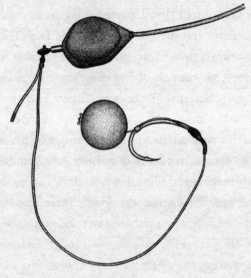

A bolt-hair rig

Safety – we've mentioned safety as a common theme through these pages because it's vital. Before John was allowed to fish alone, he had to learn to swim a length of the pool at Stockport, and if you want to fish and have thought about swimming lessons, take them. Never wade so deep that you are scaring yourself. Wade with a staff that acts as a third leg and gives you better grip. Wading or not, a modern, automatic life-preserver is a potentially invaluable investment. Take special care on steep, muddy banks like you find on the Wye; a rope tied to a tree and your waist is not going OTT. Take care in extremes of temperature. Take a charged phone. Let folks know when you are expecting to be home. Don't get yourself paranoid, but be aware of every possible hazard out there – especially when an imitation fly is whizzing about (wear glasses!). THINK! You're fishing on a steep bank in the late summer and you see wasps aplenty. There'll be a hole in that bank which will lead to a nest and possible wasp action, so move. One example of very many when it pays to have eyes wide open. Safety is all about experience and common sense most of the time, and to show we are not being all smart about this we'll admit to our own mistakes. Many years ago John took four friends out onto Loch Arkaig in Scotland in a slightly leaky boat designed to hold three adults. There was no bailing can. The engine was dodgy. There was no spare fuel can aboard. There was only one oar and no rowlocks. No one had a life jacket. The morning was rough and the wind was forecast to get to gale force by lunch. That's a madness no one should ever repeat.

Scales – to weigh or not to weigh, that is the question. Of course, weighing a fish never does it any favours, but if you do it quickly and to a well-thought-out routine, it can serve a purpose. Always weigh a fish in a wetted sling or even a plastic bag. Put the sling on the scales and zero them. Then take the fish from the water, weigh it safely and swiftly, and return it at once. It is nice to record personal bests and it can be useful to check on how well fish are growing, so we don't discount the practice altogether, though you won't see scales on *Gone Fishing*. (*Paul: We don't catch anything big enough.*) Whether personal bests are valid is a moot point. John used to be sniffy about them, but seeing the joy these pound and ounce milestones bring anglers he's rather changed his mind.

Shot – the weights you put on the line to cock a float and take a bait down the water column. These used to be lead in the

THE NUTS AND BOLTS OF WHAT WE DO

good old days, but now are made of fairly useless and very expensive substitute materials. Big shot are called SSGs or even double SSGs, and we use them a lot when river fishing. Then we have AAAs, BBs, No1s in order of decreasing size. All these are useful, but you'll rarely need to go smaller than No4s in usual fishing situations.

Sightfishing/Strategising – we've lumped these together, as they are largely the same thing. Actually SEEING fish in water clear enough to do so gives you a head start, naturally enough. At least you know there are fish in front of you, and that should give you confidence. If you see fish, you can watch what they are doing, if they are feeding and where they are going. The strategy bit comes in when you make a plan to catch the fish you are sighting. Easy! You might decide, for example, to put down a carpet of bait and wait for a fish to start feeding on top of it. Or you might flick a worm just ahead of a slow-moving carp. Exciting. Imaginative. Way to go when you get better at the real basics.

As you get better, all manner of sightings will begin to make sense. You might see that fish avoid your line as it falls through the water in front of it: plot how to nail the line on the bed with small shot so they don't see or feel it. You might realise that the fish are picking up every bit of bait but your own. Chances are it is the weight of the hook that is stopping the bait rising up as the fish suck at it. Try a hair rig or a plastic imitation of the bait that is more buoyant and counterbalances the hook weight. You'll *never* get to the end of this learning curve.

Sink and draw – as simple as you like, this pike technique. Simply attach a dead fish, a six-inch fresh roach is best, on a wire trace with two size 8 trebles. Put one treble through the deadbaits lips and the second treble deep in one of its flanks. Sometimes an SSG shot just above the wire trace helps give control and casting distance. Check the bait sinks slowly by dunking it in the margin – if it doesn't, puncture the stomach area with a knife a few times until the swim bladder is burst. Cast the bait out, let it sink and then reel it very slowly back in. Give the bait life. A couple of quick turns of the reel and then a pause. The roach will sink and then spurt into seeming life. Move the rod tip right to left so the course of the bait zig zags and it looks even more like a fish in extreme bother. Keep the retrieve going right to the bank as takes can come at your feet. If a pike hits the bait on the way in, give it slack for a few seconds, five at most, then wind down and pull steadily into your fish.

Smelling roses – it's one of those tedious cliches that appreciating flora and fauna increases your joy in the angling experience. Much as we dislike sounding as crusty and dusty as Izaak, it actually is correct, this old smelling roses lark. We've put our heads together and come up with the Whitehouse–Bailey list of fur, feather and flower you should try to recognise for your pleasure and sometimes for your fishing efficiency. Of course, this doesn't even begin to scratch the surface of what's out there, but just for starters.

Willows and **alder:** *The* trees of the riverbank. All fish love the shelter offered by these trees' submerged roots and the shade given by their spreading branches.

Marsh marigolds: Also known as kingcup, the quintessential marginal plant that blooms spring into early summer and tells us it's time for trout on mayfly and tench at dawn.

Water buttercup aka *Ranunculus lingua grandiflora*: The plant of the river gravels in the summer. Here fish spawn, whether barbel, chub or even sea lampreys. The weed is full of invertebrates and even the flowers host bees and dragonflies. It's the heart of the river. Trot a worm down those corridors between the unfurling columns of green and you are in for a treat.

Water voles: The adorable Ratty of *The Wind in the Willows* and still a resident here and there along the thick reed beds of some rivers where otters and mink haven't harried them to the ends of the earth. Look for chewed-off reed stems, little muddy runs and tiny droppings.

Swifts, swallows and **martins:** The water-skimming fighter pilots of the summer and autumn. When they are on manoeuvres you can be sure fly life is teeming and fish of all species will be on the feed.

Barn owls: The ghost of twilight, the bird that tells you the night will be mellow and a vigil by the rods until darkness at least might be worth your while.

And last but way not least, the halcyon dart of blue that is the beloved **kingfisher**. There are unlucky souls who have never seen this special bird or heard its shrill piping call, but not us, members of this lucky band of fishers.

Splash factor – John's baby, this, and all he means is that many anglers don't realise how far and fast sound travels underwater. That's why he rarely uses heavy leads or feeders in rivers and goes for 2, 3 or 4 SSG shot instead. He even spaces these shot out so they don't land with a splash, but instead a split-splat that resembles boilies or pellets hitting the water. He might overplay this particular neurosis, but you never lose a fish by being too careful.

Stick floats – there are two major float types, sticks for rivers and wagglers for still water. You use a stick when you are **trotting** (see 'T'), and you fasten it to the line top and bottom; that is, with the line attached to the tip of the float and to the bottom of the float's stem. This gives you better control when trotting. The big deal is choosing which float size for which swim. It's a compromise. You want a float big enough to cast and control, but not so big it scares the fish with its splash and shadow. Start small, and if you feel you are being swept away with no control, keep going up in float size till you feel you are fishing comfortably and efficiently. We like to place the cocking shot around midway between float and hook, often with a smaller shot, say a No1, six inches away from the bait.

A stick float

Structure – really just a pro term for any feature in the water, like a bridge's footings, a jetty, some anchor ropes, old pilings, a discarded fridge, anything that provides sanctuary from predators or fast currents. Wherever there is structure, you are likely to find fish, so look out for anything that fits the bill and fish close.

Tools – we've always tried to keep expenditure down for you and cut out the purchase of trivialities, but you will need things like baiting needles (for boilies especially), disgorgers and forceps for taking out hooks, and pliers for taking bigger hooks out of pike. Buy good ones and always have them to hand, as you don't want to be faffing about if there's a fish out of water. Scissors are always a good idea.

Touch ledgering – we've already made clear that we're not great fans of sitting back and letting a bait fish for itself. Touch ledgering is an effective, thrilling way of ledgering which is all about skill and being in the zone with the fish. We've mentioned it earlier in this book, and it really is right up there on our list of favourite fishing approaches. And it's not a black art, but a method you can pick up in hours. It's primarily a river trick and great for chub, barbel and even roach on pellets, boilies, corn and bread flake. We like to use SSGs for the weight, the number depends on the flow, the depth and the distance you are fishing, but 2 or 3 SSGs should be fine for most rivers in normal conditions and you rarely need to use more than 4 or 5, even in floods. Any more than that and the method loses its effectiveness. Space the SSGs out, starting about a foot up from the hook and at foot-long intervals thereafter. Cast out slightly upstream and keep in touch with the rig as it falls through the water and settles a little down-stream of you. Point your rod tip (an Avon-type rod is ideal) directly to where you think your bait is lying and hold it steady about a foot or two above the surface. Hold the line between reel and butt ring in your fingers. There are several ways of doing this, but I like to feel the line lodged over my forefinger. Experiment till you find what works for you. Bites can come in many guises. You might just get a full-blooded pull. It might be a steady build-up of pressure you feel, or a quick but hard snatch. Sometimes you get short but definite indications that are too quick to strike at, and these are generally fish picking up the bait and testing it. Yes, the approach is *that* sensitive. One thing is sure: when you do get a take to strike at, your instincts will kick in and your strike will be instant and generally rewarded with a hooked fish. There are huge advantages with touch ledgering over the normal stuff. You can move swims more easily, as you don't have rod-rests, and as it is done best standing up, often in the river, there's no faff with chairs. Often you feel a rasping type of sensation on the line. It's hard to put into words, but you know it is something down there that's alive; in fact, it is fish rubbing their bodies on the submerged line. This tells you there are fish in front of you and to change baits or hook sizes till you get a full-on bite. And, of course, you are not relying on a visual indication, so it's a great method in low-light conditions.

Trolling – there are many ways of using artificial lures. You can spin, for example, simply casting out a spoon and reeling it back in with a bit of life and imagination imparted as you do so. On very big waters like Scottish lochs, trolling is an effective method for big predators like pike, perch and trout. You simply tow two or three spoons, plugs or big spinners behind your boat and wait for a fish to intercept them. Boring? Not really. You need to captain the boat proficiently. You need to use the right lures and work them at the right depth and speed. You need to avoid tangles and snagging on the bed or on projecting rocks. Often the weather will be vile and you really do need your wits about you. And you can always take your binoculars and, if nothing else, watch stags on the hill or eagles soaring high.

Trotting – you trot a stick float down a river to catch roach, dace, chub, barbel and perch with a moving bait, commonly maggots, corn or bread flake (see diagram below). It's a magi-

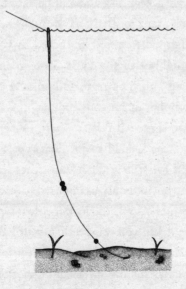

Trotting

cal way to fish and highly efficient once you master the simple basics. Trot short, ten-yard runs to begin with, and as you get more experienced you can work a float ninety yards – if you can see it. It's also a good idea to trot slower water at first, as very quick water brings problems of its own. **Mending the line** is a big deal – you lift the rod and simply straighten out loops in the line between your rod tip and the float. If you don't do this, the loops will get bigger and bigger, and two things happen. First, the float gets pulled all over the place, and second, if you do get a bite, there will be too much loose line to set the hook.

Umbrellas – now, we don't use them simply because we're not ones for setting up our stall and sitting in one place all day. We believe that if we buy good rainproof hats, coats and overtrousers we'll keep dry and still be able to move around easily when we want to. That's us! It might be that you want to fish matches, when you'll be rooted to your peg for the duration. It might be that you just like to get to a good swim, set up, cast out and then relax and enjoy a peaceful day. It rains? You simply put up your brolly and watch the ducks. Who are we to argue with that?

Waggler floats – the second major float type is the wagglers, generally used for still-water fishing. These are attached to the line by the bottom only, as opposed to stick floats, which are fixed top and bottom. You lock the float with plastic stops or shot at the depth you want to fish your bait, generally on the river or lake bed. John likes to use shot to cock the float, and on still waters there is no need for shot on the line, as it falls through the water column. If you are using pellets or

Using a waggler float

boilies there is enough weight to take the bait to the bottom and to hold it there in all but rough conditions, when a BB shot might be used close to the hook. Finally, if there is no shot on the line, it hangs limp, as soft and giving as a strand of loose weed. If a fish brushes along it, the feel is natural and holds little menace. These niceties might sound overly fastidious, but they are not. Very often, the slightest tweak can make the biggest difference.

Watercraft – the core of good fishing practice. You look at any water and learn to read it. You look for features, structure, what currents are doing, where the wind is blowing, for islands, weeds, reeds, for incoming streams, overhanging trees, water-submerged tree roots, for flies hatching, jumping fish, rolling fish, bubbling fish, for coloured water where fish are

feeding, shallows, gravel bars, for a million things that give you clues and are all around. Regard every water as a great and wonderful painting waiting to be deciphered. Watercraft is all about putting the fish and water first, and the bait and methodology second. This is what the thinking fisher is all about, and it is what makes the sport the fantastic learning experience we all enjoy.

Weather – fish are breathing barometers as highly tuned to weather conditions as any living creature. Most fish feed at some time every day apart from when there is a disastrous drop in temperatures in the winter and those days when the wind blows from the north and especially the east. Avoid both if you can. Stable conditions are great, winter and summer, and a rise or fall in air pressure can jolt lethargic fish into feeding. For example, if it is very hot in the summer, a spell of cooler, wetter weather with westerly winds can be superb. Or, conversely, if it's cold and wet, warmer winds from the south and a rise in pressure can drive fish to a feeding frenzy.

There are many anglers who obsess about lunar phases, but we have truthfully never found a full moon better or worse than a new one. We just don't know about lunar phases, and we admit this happily because mystery is at the core of angling's magic. Yes, we are good, experienced anglers, we like to think, but in the great scheme of things our knowledge is limited. Even at our age, we feel we are only beginning our angling journey and we'll be learning more about fish until the day we die. That is fishing. It can take you from the cradle (almost) to the grave, and every session along the way is filled with enlightenment and endless elements of enjoyment.